MEDIA AND COMMUNICATIONS - TECHNOLOGIES, POLICIES AND CHALLENGES

INTELLIGIBILITY RESEARCH AND COMMUNICATION ISSUES IN EMERGENCY SITUATIONS

MEDIA AND COMMUNICATIONS – TECHNOLOGIES, POLICIES AND CHALLENGES

Additional books in this series can be found on Nova's website at:

https://www.novapublishers.com/catalog/index.php?cPath=23_29&seriesp=Media+and+Communications+-+Technologies%2C+Policies+and+Challenges

Additional E-books in this series can be found on Nova's website at:

https://www.novapublishers.com/catalog/index.php?cPath=23_29&seriespe=Media+and+Communications+-+Technologies%2C+Policies+and+Challenges

MEDIA AND COMMUNICATIONS - TECHNOLOGIES, POLICIES AND CHALLENGES

INTELLIGIBILITY RESEARCH AND COMMUNICATION ISSUES IN EMERGENCY SITUATIONS

SAMUEL A. FLETCHER
EDITOR

Nova Science Publishers, Inc.
New York

LIBRARY OF CONGRESS CATALOGING-IN-PUBLICATION DATA

Intelligibility research and communication issues in emergency situations / editor, Samuel A. Fletcher.
p. cm.
Includes index.
ISBN 978-1-61668-634-5 (hardcover)
1. Public safety radio service. 2. Emergency communication systems. 3. Speech, Intelligibility of. I. Fletcher, Samuel A.
TK6570.P8I58 2010
384.5'34--dc22

2010012234

Published by Nova Science Publishers, Inc. † New York

CONTENTS

Preface **vii**

Chapter 1 Intelligibility of Selected Radio Systems in the Presence of Fireground Noise: Test Plan and Results **1**
J. David Atkinson and A. Andrew Catellier

Chapter 2 Relationships between Intelligibility, Speaker Identification, and the Detection of Dramatized Urgency **99**
A. Andrew Catellier and D. Stephen Voran

Chapter 3 Emergency Communications: Vulnerabilities Remain and Limited Collaboration and Monitoring Hamper Federal Efforts **127**
U.S. Government Accountability Office

Index **197**

PREFACE

The September 11, 2001, terrorist attacks and response to Hurricane Katrina in 2005 exposed the severe consequences of breakdowns in emergency communications used by first responders. Failures in emergency communications resulted in numerous lost lives and exacerbated already challenging situations. These past events have increased focus on the need to enhance emergency communications to respond more effectively to future catastrophic disasters. Effective response to catastrophic disasters will require that first responders - law enforcement personnel, firefighters, and others first on the scene - have reliable communication systems, including supporting infrastructure, facilities and staff. This book focuses on the issues and vulnerabilities to emergency communications systems, the federal assistance available and other challenges with these efforts.

Chapter 1 - Project 25 (P25) is an initiative to develop a standardized digital radio and promote interoperability among digital land mobile radio (LMR) systems. The Association of Public Safety Communications Officials (APCO) and the Telecommunications Industry Association (TIA) cooperated to form this initiative in 1988. An integral part of the initiative, the digital voice coder/decoder pair (or vocoder for short) was selected in 1992, based on the results of several tests. This vocoder is named for the technology it uses, Improved Multi-Band Excitation (IMBE™), and was standardized by the TIA as TIA-102.BABA.

The performance of the IMBE vocoder has proven to be problematic in tactical fireground communications, where considerable background noise may be present. Several fire agencies have brought this problem to public attention (Boise, Fairfax, Littleton, and Phoenix Fire). The Digital Problem Working Group (DPWG) was formed by the International Association of Fire Chiefs (IAFC) in response to the problem, and is designed to provide input to testing procedures. The purpose of DPWG is to define the problem, identify potential solutions, and recommend best practices that could mitigate issues identified by first responders.

Chapter 2 - The systems used for public safety speech communications must be intelligible. It is also desirable that they transmit secondary information, such as the attributes of a speaker's voice. This secondary information can allow a user to identify the speaker and his or her emotional state. Testing speech communications systems for the delivery of intelligible speech is common. Testing for human perception of the delivery of this secondary information is less common, though some prior work has been done. Building on this prior work, we describe a set of controlled laboratory listening experiments. These experiments characterize the relationships between speech intelligibility, speaker identification, and the detection of dramatized urgency in a speaker's voice across a range of simulated speech processing conditions. The experiment results indicate that for the speech processing conditions considered here, detection of dramatized urgency is the most robust property, speaker identification is less robust, and speech intelligibility is the least robust.

Chapter 3 - Emergency communications breakdowns undermined response efforts during terrorist attacks in 2001 and Hurricane Katrina in 2005. In response, federal agencies like the Department of Homeland Security (DHS) and Federal Communications Commission (FCC) have increased efforts to enhance emergency communications. This requested report identifies (1) vulnerabilities, if any, to emergency communications systems; (2) federal assistance available or planned to first responders for addressing vulnerabilities or enhancing emergency communications; and (3) challenges, if any, with federal emergency communications efforts. GAO developed six catastrophic disaster case studies, reviewed agency documents, and interviewed public and private sector officials at the national, state, and local levels.

In: Intelligibility Research and Communication Issues… ISBN: 978-1-61668-634-5
Editor: Samuel A. Fletcher

Chapter 1

INTELLIGIBILITY OF SELECTED RADIO SYSTEMS IN THE PRESENCE OF FIREGROUND NOISE: TEST PLAN AND RESULTS*

J. David Atkinson and A. Andrew Catellier

ABBREVIATIONS/ACRONYMS/GLOSSARY

Codec	Combination of an encoder and decoder in series (encoder/decoder).
Coder	Same as "encoder".
Decoder	Device for translation of a signal from a digital representation into an analog format. For the purposes of this document, this is a device compatible with TIA102.BABA [1].
df	Degrees of freedom for ANOVA or Newman-Keuls calculations.
Encoder	Device for converting an analog signal into a digital representation. For the purposes of this document, this is a device compatible with TIA-102.BABA[1].
Fireground	The fire service term for an active fire scene.
FM	Frequency modulation.
HATS	Head and torso simulator.
IMBE™	Improved Multi-Band Excitation. A speech encoder which divides the audio spectrum into bands and generates speech model parameters based upon these spectral bands. In the speech decoder, the model parameters are used to synthesize speech for each of the spectral bands.
LLR	Log likelihood ratio.
LRP	Lip reference point of the HATS.
MRT	Modified rhyme test as defined in [3].

* This is an edited, reformatted and augmented version of a U. S. Department of Commerce publication dated June 2008.

PASS	Personal alert safety system, which emits a signal in the event that the user becomes incapacitated or needs assistance.
PCM	Pulse coded modulation, a logarithmically companded and 64 kb/s encoded representation of speech.
Pink Noise	Acoustic noise that has equal power per octave band as frequency increases.
QFA	Full-rate baseline codec with no modulation (7200 bps).
QFB	Enhanced full-rate codec with no modulation (7200 bps).
RF	Radio frequency.
Vocoder	Voice encoder / decoder.

EXECUTIVE SUMMARY

Project 25 (P25) is an initiative to develop a standardized digital radio and promote interoperability among digital land mobile radio (LMR) systems. The Association of Public Safety Communications Officials (APCO) and the Telecommunications Industry Association (TIA) cooperated to form this initiative in 1988. An integral part of the initiative, the digital voice coder/decoder pair (or vocoder for short) was selected in 1992, based on the results of several tests. This vocoder is named for the technology it uses, Improved Multi-Band Excitation (IMBE™), and was standardized by the TIA as TIA-102.BABA.

The performance of the IMBE vocoder has proven to be problematic in tactical fireground communications, where considerable background noise may be present. Several fire agencies have brought this problem to public attention (Boise, Fairfax, Littleton, and Phoenix Fire). The Digital Problem Working Group (DPWG) was formed by the International Association of Fire Chiefs (IAFC) in response to the problem, and is designed to provide input to testing procedures. The purpose of DPWG is to define the problem, identify potential solutions, and recommend best practices that could mitigate issues identified by first responders.

One such testing procedure has been conducted. The Institute for Telecommunication Sciences in Boulder, Colorado developed the test plan in conjunction with the DPWG Testing Subcommittee. The goal of this experiment was to measure the intelligibility of communication systems operating in high acoustic noise environments typical of those encountered by firefighters. The three primary communication systems consist of:

- 25 kHz analog FM radio pair (a.k.a. 25 kHz Analog)
- Baseline full rate IMBE vocoder radio pair (a.k.a. P25 Full Rate)
- Enhanced full rate IMBE vocoder radio pair (a.k.a. P25 Enhanced Full Rate)

A modified rhyme test (MRT) was used to evaluate each communication system in each of nine environmental noise conditions (described below). The MRT method is more completely described in ANSI S3.2, and is required in the evaluation of self-contained breathing apparatus (SCBA) by the National Fire Protection Association (NFPA) 198 1-2007 standard.

In short, a basic MRT is a test where speakers are instructed to say "Please select the word," followed by one of any number of words that "rhyme" on either the first or last consonant (e.g., one group of six words includes bed, led, fed, red, wed, and shed). These utterances are recorded, and later played back to test subjects, or listeners. The listeners are asked to select the word they heard at the end of the sentence. There are specific methods used to interpret this data, due to the situation where the listener has a limited set of possible answers to choose from, advance knowledge of the majority of the incoming sentence, as well as other constructs of the testing environment.

In this test, the recorded utterances were processed using the vocoders on the aforementioned radio pairs. That processing took place while the radio pair was in the presence of one of the following nine environmental noise conditions:

- No background noise, no mask (or the Clean condition)
- Fire truck pump panel, no mask
- Mask with no background noise
- Two Personal Alert Safety System (PASS) alarms, with mask
- In-mask low-air alarm
- Rotary saw cutting metal garage door, with mask
- Chainsaw cutting wood, with mask
- 2 1/2" hose with fog nozzle, with mask
- Rotary saw cutting metal garage door, with amplified mask

Those processed files were played back to listeners in a sound-isolated room. For the duration of the test, the room was filled with a field of pink noise. This test also evaluated a 12.5 kHz analog FM radio pair on three of the nine environmental noise conditions. The inclusion of this radio pair was to evaluate the intelligibility of this communication system, which meets an FCC mandate requiring narrowband devices.

After 30 listeners participated in the test, the results were processed and some interesting conclusions can be drawn. The performance of the 25 kHz Analog system was either statistically similar to or better than the P25 systems for all environments. Four of the nine environments were too difficult for intelligible communication using all tested systems (i.e., less than 10% intelligibility). The 12.5 kHz Analog system was statistically similar to the 25 kHz Analog system.

This testing examined fire safety equipment used in conjunction with these systems. In the case of voice transmissions while using a mask, there was a significant degradation in intelligibility for the P25 vocoders. The PASS alarm, designed to augment safety of first responders, significantly degraded intelligibility of both P25 vocoders. The low air alarm was effectively too difficult a noise environment for all of the tested communications systems, but it is worth noting that the 25 kHz Analog system and the P25 Enhanced Full Rate system preserved the noise characteristics of the low air alarm sufficiently well that a listener could determine what type of alarm was sounding.

While this information may be useful to those planning to purchase and deploy new radio systems for their agencies, these results should not be the sole source of information. Other decision factors should include which agencies are involved, their current assets, their operating procedures, policies, and budget plans, spectrum availability in their locale, State

Communications Interoperability Plans (SCIPs), and other, more specific situations that are beyond the scope of this document.

This report describes an experiment conducted to measure the intelligibility of selected radio communication systems when those systems are employed in highbackground-noise environments experienced by firefighters. The test plan for a Modified Rhyme Test (MRT) is detailed, including requirements for source material preparation and listening test conduct. Finally, the results of the test are presented, along with the data analysis. The results indicate that in some environments analog radios performed better than digital radios, and in some environments no radios performed well. This information should be considered whenever an agency is preparing to purchase and deploy a new communications system.

Keywords: intelligibility; Project 25; vocoder; modified rhyme test; noise; analog FM; land mobile radio; LMR; public safety; fire service

1. Introduction

Project 25 was initiated in 1998 as a cooperative agreement between the Association of Public Safety Communications Officials (APCO) and the Telecommunications Industry Association (TIA). The purpose of Project 25 was to develop a standardized digital radio to promote interoperability among digital land mobile radio (LMR) systems. One element of the Project 25 suite, the vocoder, was selected in 1992. At that time, several tests were conducted by the Project 25 committees to ensure that the best available vocoder was selected. The selected vocoder, known as the Improved Multi-Band Excitation (IMBE^{TM}, a trademark of Digital Voice Systems Inc.), was standardized by TIA as TIA-102.BABA.

As P25 networks were deployed, it became evident that certain noisy environments in which public safety must communicate are problematic for the vocoder. This has appeared most consistently in tactical fireground communications, and the issue has been raised to the national level by agencies such as Boise [Idaho] Fire, Fairfax [Virginia] Fire, Littleton [Colorado] Fire, and Phoenix [Arizona] Fire. In response to this issue, the International Association of Fire Chiefs (IAFC) created a Digital Problem Working Group (DPWG) to provide expert input into the testing to specifically identify the problem and potential solutions as well as to develop best practices that could mitigate some of the problematic issues in the communications environment.

This document provides the test plan and results of the first test undertaken at the request of the DPWG. The test was designed to compare intelligibility of communication systems being used in fireground noise situations. The test plan was developed in conjunction with the Testing Subcommittee of the DPWG and the testing was conducted at the Institute for Telecommunication Sciences in Boulder, Colorado. This document describes the evaluation procedure used to characterize the response of digital voice coding technology to public safety noise environments, as well as the results of that procedure. The report specifically focuses on those background noise environments experienced in the fireground. Understanding the characteristics of a digital voice coder is essential to enabling effective communication in the environment in which public safety personnel must operate.

The evaluation procedure is designed to give a fair characterization of the communication system. This is accomplished through the examination of the system performance in a subjective listening test in which the relative performance among the enhancements is measured in a quantitative way.

The purpose of the subjective listening test is to evaluate communication systems under a variety of operating conditions. The operating conditions were chosen to be representative of those experienced in a fireground environment. Only a limited number of operating conditions are tested. A test of all possible operating conditions would be too unwieldy to conduct.

Section 4 describes the communication systems (and their respective vocoders) that are the subject of this test. Section 2 also discusses any background noise for the test conditions. Section 3 discusses the speech database used in the testing and comparison. Section 4 contains the details of the testing procedure as well as the speech database used for the testing. Section 5 discusses the design of the listening tests themselves, including the analysis of results. Section 6 describes the disclosure of the test results. Lastly, appendices are included which contain additional detail.

1.1. Scope

This document specifies the procedures to be employed to characterize the behavior of TIA 102.BABA compatible speech codecs in environmental noise conditions. The original baseline speech codec from 1992 is the IMBE described in TIA 102.BABA, Project 25 Vocoder Description. The IMBE speech codec is used to digitally encode the speech signal and provide forward error control for transmission at a data rate of 7200 b/s. The baseline speech codec is defined as the reference codec for the Project 25 MOS Conformance Test [2].

The testing of the codecs is performed using subjective listening tests that judge overall speech intelligibility [3].

Where possible, this study compares systems incorporating the speech codecs with those using analog FM. The baseline speech codec is based upon the Digital Voice Systems Incorporated implementation of the IMBE algorithm on the VC 20 Project 25 hardware card, or the equivalent software version. The software version is of primary interest in this test. The experiment compares the coding mechanisms with various channel conditions as might occur on a land mobile radio channel.

1.2. Overview

Speech coders are tested by comparing their performance with a reference implementation such as analog FM. The test evaluates the intelligibility of communication systems using different vocoder technologies, including reference communication systems. The test results are then compared against the reference. There are four communication systems (with 4 different coding technologies) in this test. These mechanisms are listed below, and they are also described in Section 4.

System 1	25 kHz analog FM radio pair.
System 2	Radio pair implementing the baseline full rate IMBE vocoder. This is referred to as the “P25 Full Rate” throughout this report.
System 3	Radio pair implementing the enhanced full rate. This is referred to as the “P25 Enhanced Full Rate” throughout this report.
System 4	The experiment will also evaluate the viability of 12.5 kHz analog FM radio in three of the background noise conditions.

The inclusion of the 12.5 kHz analog FM system provides an indication of whether or not that might be a viable alternative that meets the narrowbanding mandate of the FCC while still providing a required level of intelligibility to the communication system user.

The intelligibility of each communication system is subjectively rated in each environmental noise condition specified.

The intelligibility of a communication system can be difficult to quantify since it is a subjective issue, relying on humans to be able to discern words. This performance evaluation relies upon subjective testing using a panel of listeners who listen to speech passing through a system and attempt to understand what was spoken. Since discernment of listeners may vary, the results from a number of listeners are obtained and averaged to obtain an overall score.

To evaluate the intelligibility, it is necessary to conduct an experiment in a controlled manner so that unintentional variation in the scoring is avoided. The purpose of the testing is to provide understanding of the behavior of communication systems in noisy environments and to determine differences in performance among the different speech coding mechanisms. The confidence we have that any apparent differences in performance are due to communication system effects and not random statistical variation depends upon how well we prevent differences from occurring in the testing. The statistical controls for the experiment and the analysis are given in Section 5. The listening test evaluates the communication systems under operating conditions, particularly different acoustic background noise conditions on the transmitting end of the communication path. The acoustic background noise conditions represent some noisy acoustic environments encountered by land mobile radio users in the fireground. This experiment uses the following noise environments on the transmitting end of the communication path:

Environment 1.	No background noise, no mask (referred to as the Clean condition)
Environment 2.	Fire truck pump panel, no mask
Environment 3.	Mask with no background noise
Environment 4.	Two Personal Alert Safety System (PASS) alarms, with mask
Environment 5.	In-mask low air alarm
Environment 6.	Rotary saw cutting metal garage door, with mask
Environment 7.	Chainsaw cutting wood, with mask
Environment 8.	21/2” hose with fog nozzle, with mask
Environment 9.	Rotary saw cutting metal garage door, with amplified mask

Section 4.3 provides additional information about the background noise.

The system implementations are executed in hardware and recorded to generate test material. The test material consists of a series of computer files in WAV format, and the output files are in the same format for use in the listening test.

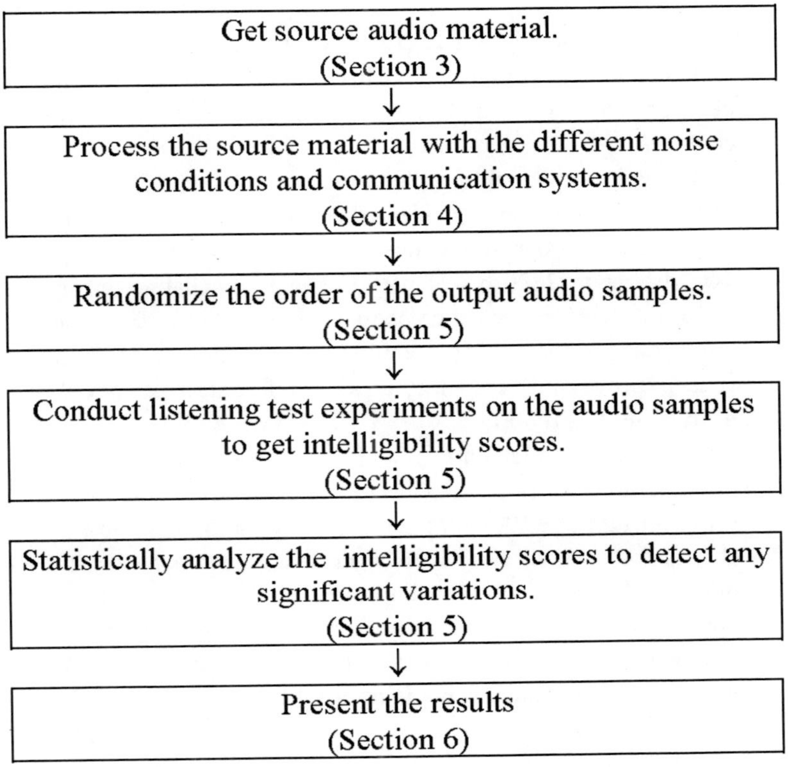

Figure 1. Test plan overview.

The overall plan of the test is outlined in Figure 1. The test begins with the source audio material. There are several conditions that the source material must satisfy, and these are covered in Section 3. The source audio material is then passed through the different communication systems, with different operating conditions, to produce numerous output audio files. This procedure is given in Section 4. The output audio files are then randomized in order to provide samples suitable for a listening test. The randomization step, together with the listening test and the analysis, is described in Section 5. The result of the test is then presented in Section 6, which describes a spreadsheet for this analysis.

2. Communication Systems and Noise Environments

This section describes the speech coding mechanisms evaluated in this test in a general way. The acoustic environments for the test are also described.

2.1. Communication Systems

There are four communication systems defined for this test:

System 1. 25 kHz analog FM radio pair. This is the reference standard for fire agencies because it is the technology that they have been using since the mid-20th century.

System 2. Radio pair implementing the baseline full rate IMBE vocoder (QFA)[2], equivalent to the vocoder selected by Project 25 in 1992. This is described as a Full Rate vocoder. It operates at a gross bit rate of 7200 b/s, a net bit rate of 4400 b/s, with 2800 b/s of parity checks for channel error correction. This includes speech processing enhancements that have been incorporated by the radio manufacturer. This combination of a baseline vocoder with optional speech processing enhancements is referred to as the "P25 Full Rate" throughout this report.

System 3. Radio pair implementing the enhanced full rate vocoder (QFB), interoperable with the baseline full rate vocoder. This vocoder operates at a gross bit rate of 7200 b/s, a net bit rate of 4400 b/s, with 2800 b/s of parity checks for channel error correction. It includes non-essential improvements developed since 1992 for improved audio quality built directly into the vocoder as well as any enhancements that may have been incorporated by the manufacturer. This combination of the enhanced vocoder with any manufacturer enhancements is referred to as the "P2 5 Enhanced Full Rate" throughout this report.

System 4. 12.5 kHz analog FM radio pair limited to three noise conditions. This is a system that could provide similar characteristics to 25 kHz Analog FM but still meet the impending 12.5 kHz narrowbanding requirements.

The radio frequency (RF) communication path between transmit and receive units will be an ideal (cabled) path.

2.2. Acoustic Environments

The experiment tests the performance of the communication systems (and their respective vocoders) with background noise mixed in with the speech at the transmitting radio for a specific signal to noise ratio. There are nine acoustic environments to be evaluated in this experiment: no background noise (no mask), fire truck pump panel (no mask), mask with no background noise, two PASS alarms (with mask), in-mask low air alarm, rotary saw cutting metal garage door (with mask), chainsaw cutting wood (with mask), a 21/2" hose with a fog nozzle (with mask), and rotary saw cutting a metal garage door (with amplified mask).

The noise conditions represent common acoustic noise environments where a user might be transmitting, and were chosen as representative by a panel of fire practitioners participating in the IAFC testing committee. The samples were taken as high-quality digital recordings made at an agency training facility. The signal-to-noise ratios (SNRs) are chosen to approximate sound level conditions that were measured in the application environment.

The power spectral densities (PSDs) and spectrograms of the six background noises are given in Figure 2 to Figure 7.

The PSDs in Figures 2-7 and the Acoustic Path Loss in Figure 8 were computed using Matlab™ mathematical analysis software. The computation method (Welch)[4], number of elements in the transform (Nfft), and the size and shape of the computational window (Hamming [5], 1024) were parameters provided to the Matlab algorithms.

The spectrograms in Figures 2-7 are plots of frequency content (on the vertical axis) versus time (on the horizontal axis).

Together the PSDs and spectrograms provide an indication that the noise environments cover a wide range of frequency characteristics, impulse characteristics, and amplitudes.

The mask acoustic transfer characteristic is given in Figure 8. There is notable signal loss in the 1,500 – 3,000 Hz range. This is significant because this range of frequencies is significant to intelligibility.

2.3. Test Conditions

The available systems combine with the noise environments to provide the array of conditions for the experiment. Table 1 shows which environments are used with which communication systems. Appendix A contains a full list of conditions for the experiment.

Table 1. Combination of Noise Environments and Communication Systems.

Environment	SNR (dB)	25 kHz Ana FM	P25 Full Rate	P25 Enh Full Rate	12.5 kHz Ana FM
Clean		X	X	X	X
Fire truck pump panel, no mask	4	X	X	X	
Mask with no noise		X	X	X	X
Two PASS Alarms, mask	-2[3]	X	X	X	
In-mask low air alarm, mask	15[3]	X	X	X	
Rotary saw cutting metal garage door, mask	4[3]	X	X	X	X
Chainsaw cutting wood, mask	5[3]	X	X	X	
21/2" fog nozzle, mask	9[3]	X	X	X	
Rotary saw cutting metal garage door, amplified mask	4	X	X	X	

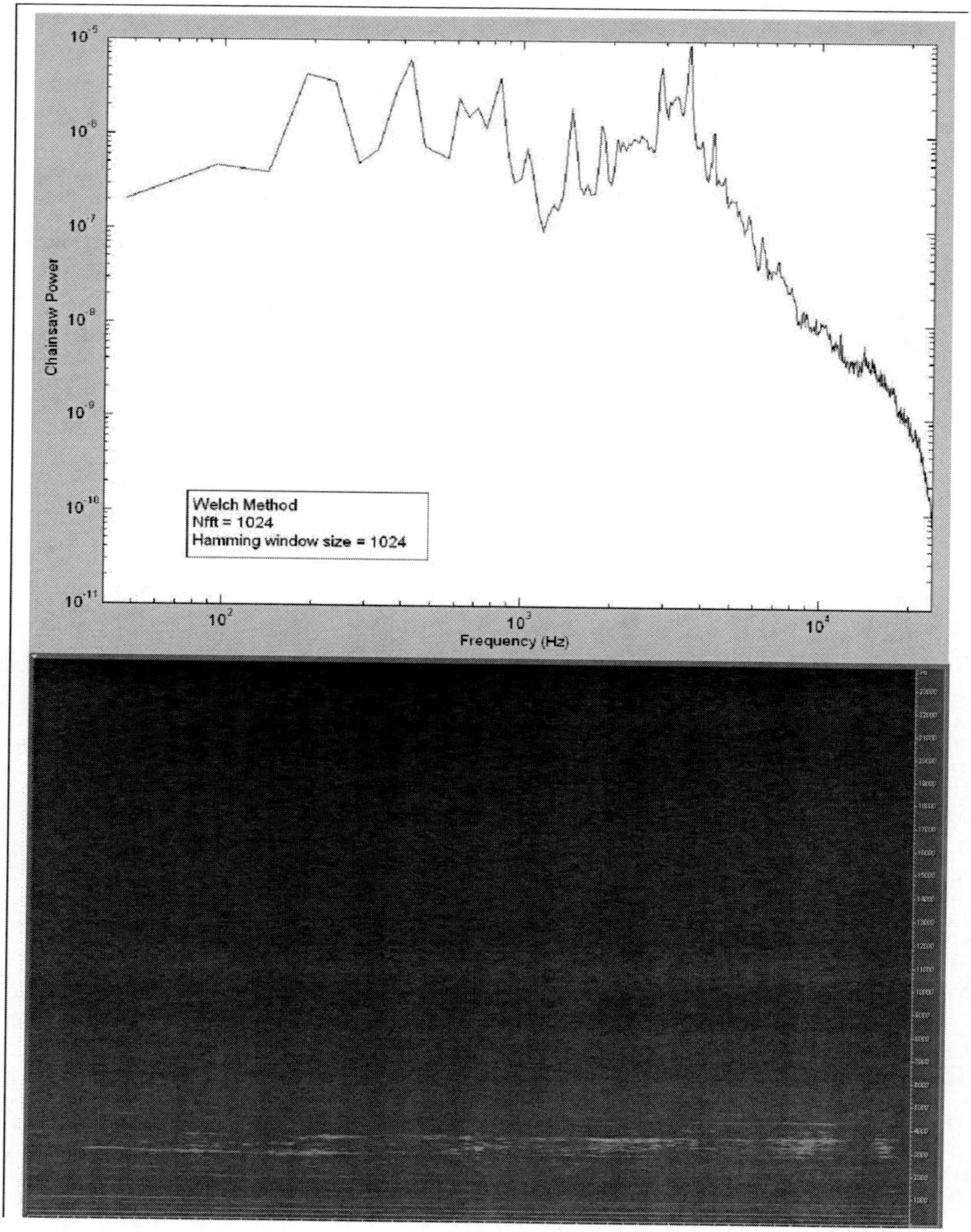

Figure 2. PSD and spectrogram of a chainsaw cutting a wood roof.

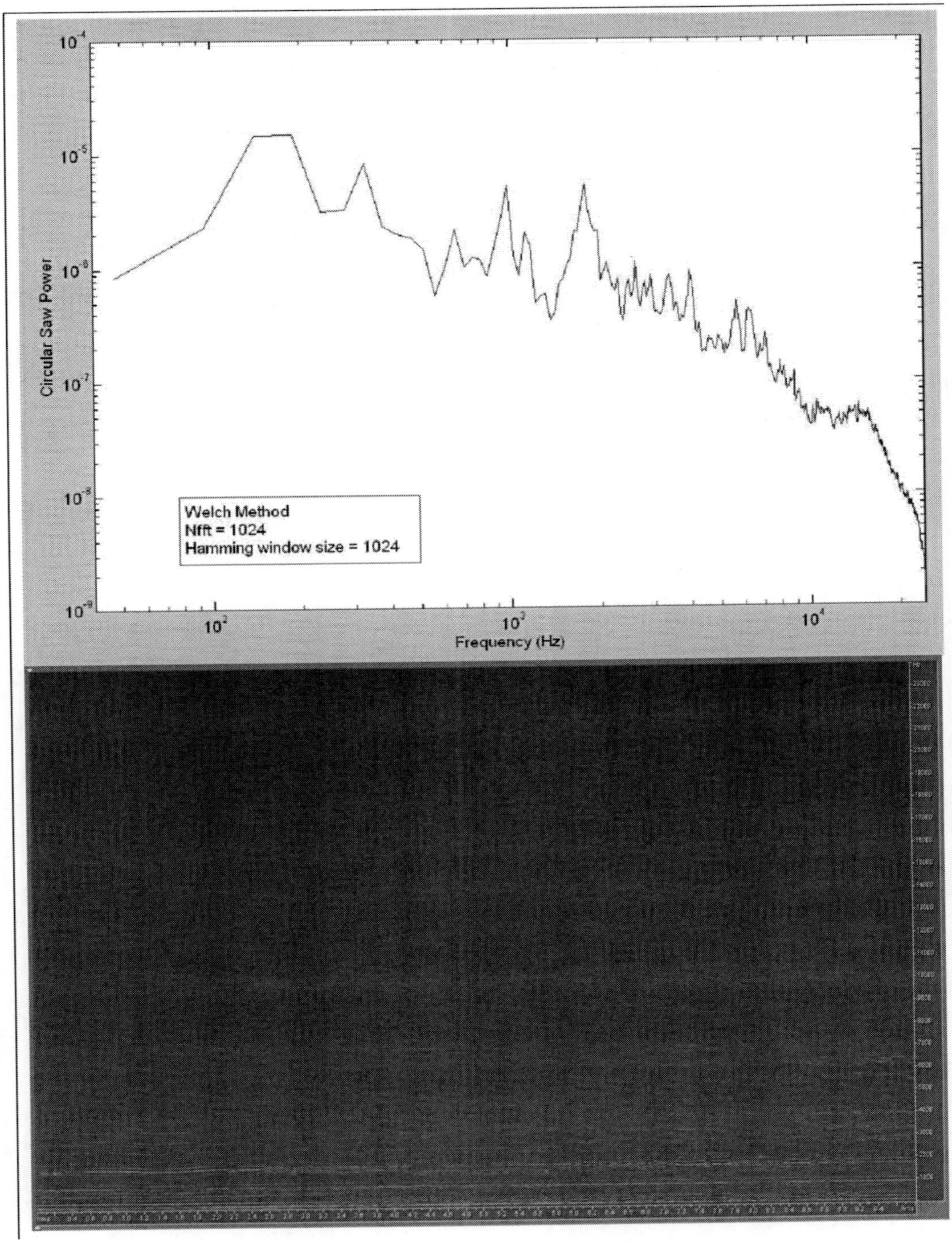

Figure 3. PSD plot and spectrogram of circular saw cutting a metal garage door.

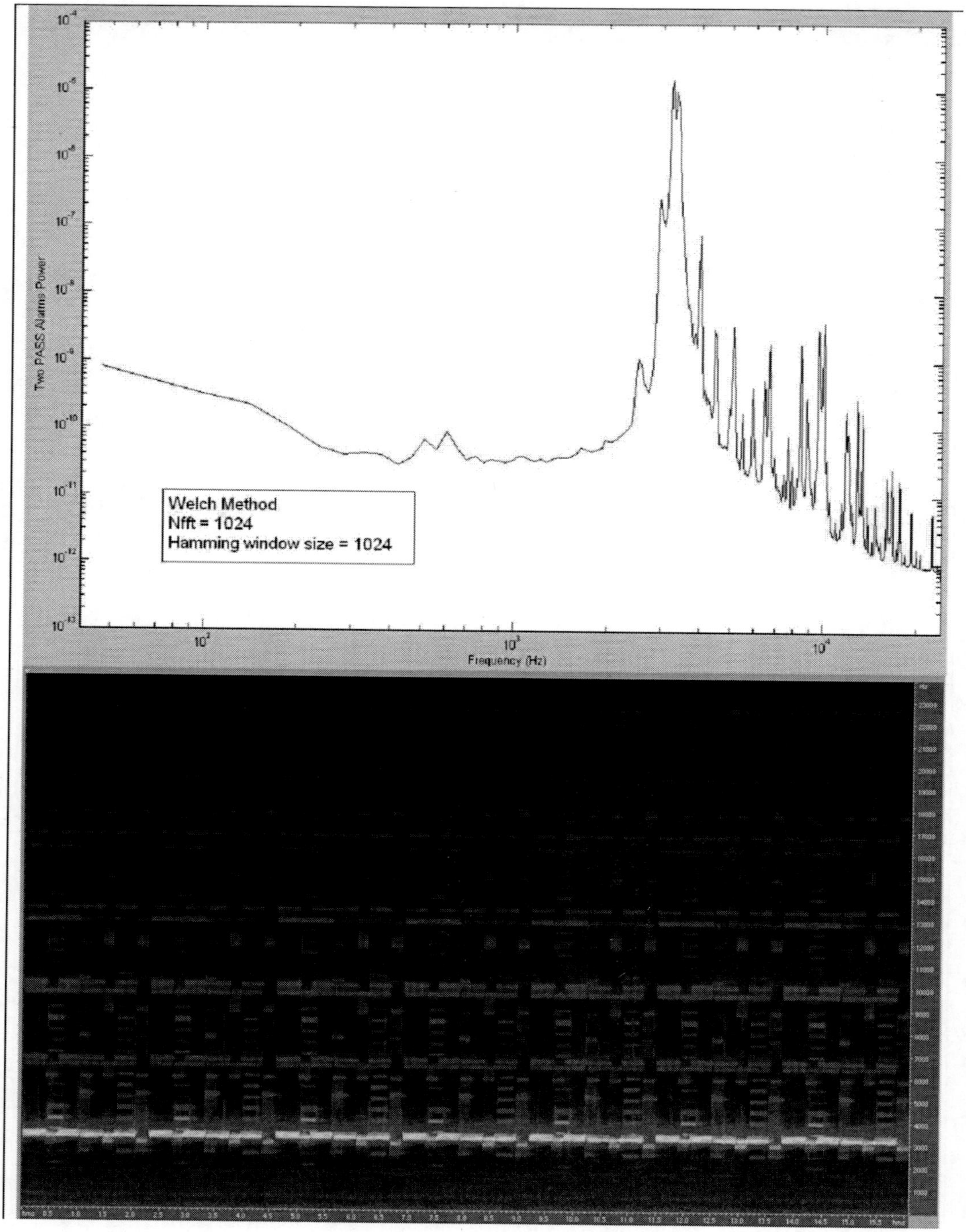

Figure 4. PSD plot and spectrogram of two PASS alarms sounding.

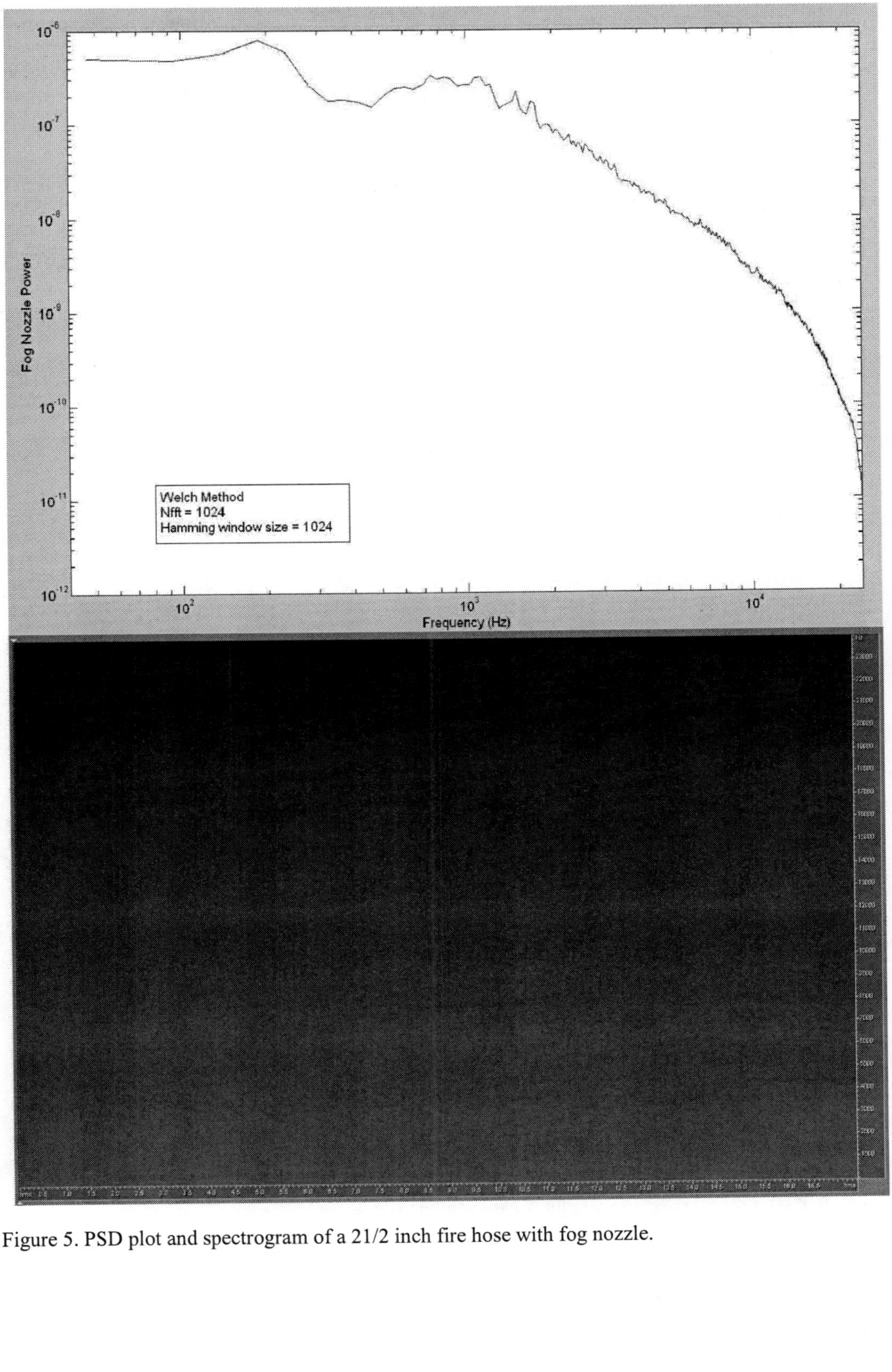

Figure 5. PSD plot and spectrogram of a 21/2 inch fire hose with fog nozzle.

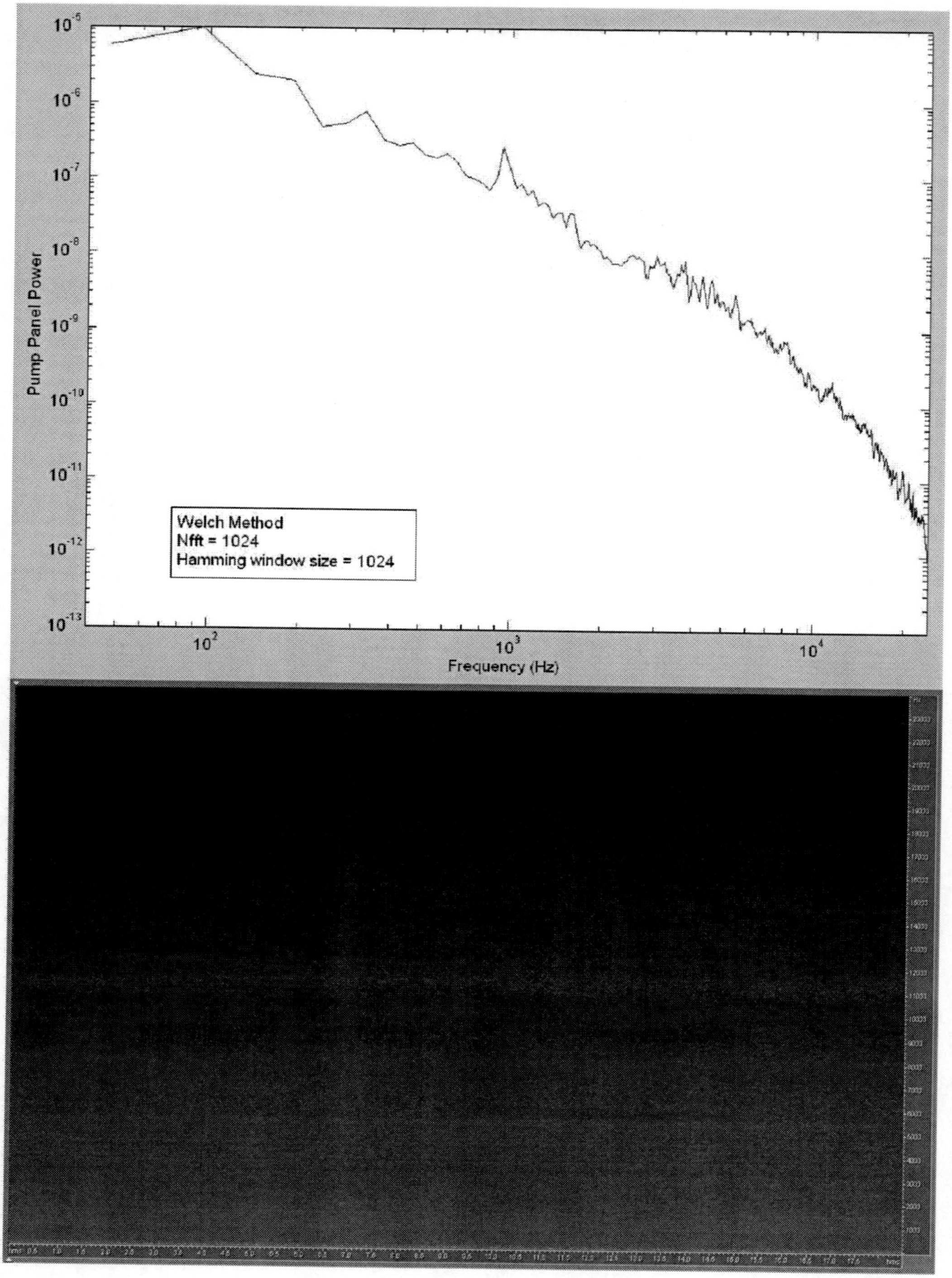

Figure 6. PSD plot and spectrogram of a fire truck pump panel.

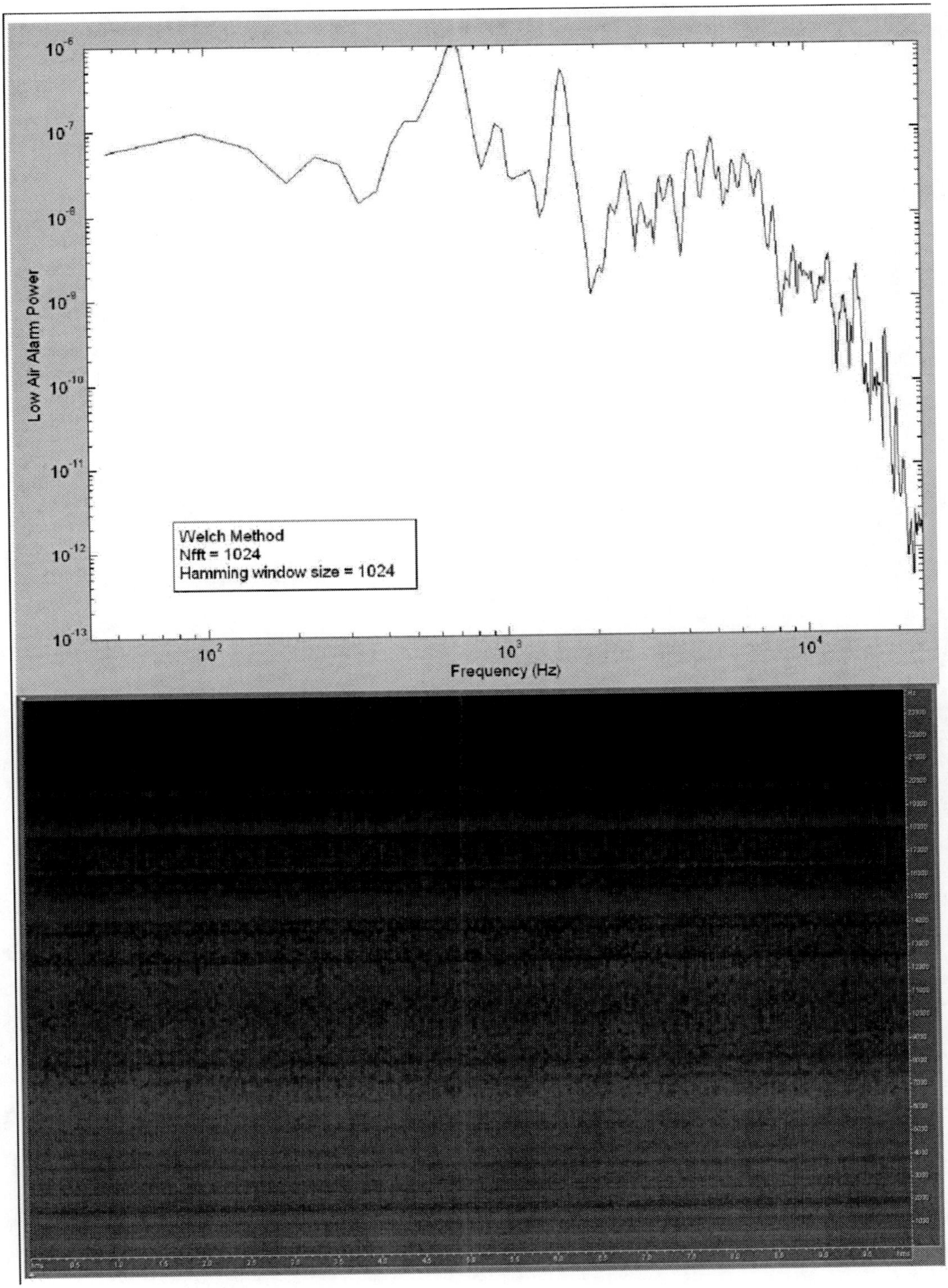

Figure 7. PSD plot and spectrogram of in-mask low air alarm.

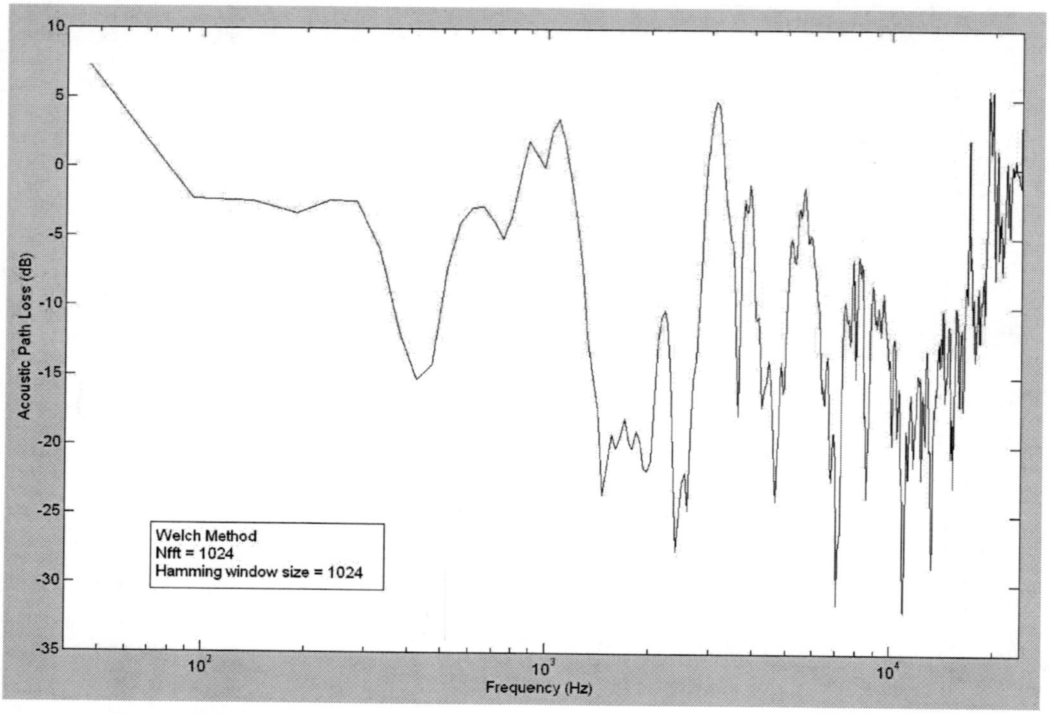

Figure 8. Acoustic path loss through SCBA mask.

3. Source Speech Databases

The speech source material used for the test consists of spoken word lists from the Modified Rhyme Test (MRT) described in [3] (e.g., one group of six rhyming words includes bed, led, fed, red, wed, and shed). Source material used for the tests has a quiet acoustic background. In the experiment, acoustic background noise is added to the source material. The MRT is used for evaluation. These tests place several requirements on the speech database:

1. Large numbers of words,
2. Equalized presentation levels,
3. Sentences of equivalent content and structure.

The speech source material for the test has been recorded at ITS. This material is the MRT word list defined in [3] using the carrier sentence, “Please select the word ...” More specific information about the recording process for this source material is in Appendix C.

3.1. Speech Database Requirements

Each test condition uses the six lists of 50 words each defined in [3]. Each coding mechanism processes the same speech material under the same operating conditions. Sufficient source material is used to ensure listeners are not presented repeat material. To reduce order bias, the presentation order of the material to the listeners is randomized.

The experiment utilizes speech recorded in a quiet acoustic noise environment. The testing includes a clean condition, and seven acoustic noise environments. For the experiments, there are three male talkers and three female talkers. [4] All talkers speak the same six lists of 50 words and are required by [3] to have little or no discernable accent.

The speech material is equalized across all talkers for presentation level. The material is provided in WAV files in full audio bandwidth and dynamic range. There are no limitations on the speech material other than those imposed by the microphone and recording system.

3.2. Speech Database Levels

An important attribute of the speech database, especially for the encoding procedure, is the average power level of the speech material. The nominal power level for speech follows the recommendation defined in [1] and excerpted below. This recommendation is followed for each sentence.

"It is recommended that the analog input gain be set such that the RMS speech level under nominal input conditions is 25 dB below the saturation point of the A-to-D converter. This level (-25 dBm0) is designed to provide sufficient margin to prevent the peaks of the speech waveform from being clipped by the A-to-D converter."

A further specification is given for the measurement method to determine the average speech power level. The varying nature of speech signal amplitude presents a difficulty for accurately measuring the level using an arbitrary method. This document uses the ITU-T Recommendation P.56 [6], method B to accurately measure the active speech level.

4. Production of Processed Files

This section describes the test procedure to be followed to conduct the tests of the encoding technologies. These procedures have been designed to assist interested parties in reproducing the speech files for later scoring by a listening laboratory.

4.1. Required Elements

The production of the mixed speech and noise files requires the following elements.

1. Head and torso simulators [7][8]
2. NC-35 sound attenuated chamber
3. Representative SCBA mask with in-mask low-air alarm

4. Ability to produce environmental noise at appropriate level within the NC-35 chamber
5. Radios that can implement the coding technologies specified
5. Recording and playback hardware and software

4.2. Test Signal Preparation

The test material generation procedure is most simply described as passing 300 sentences from each talker through nine noise environments and either three or four coding technologies for each speaker as identified in Table 1. This results in 54,000 processed sentences.

For production of processed files, clean speech is played through a HATS speaking into a radio microphone while background noise is generated by loudspeakers in the attenuated chamber. Depending on the environment, there may be a SCBA mask installed on the HATS.

For the conditions with background noise, it is important that the background noise be active before the push-to-talk is initiated to avoid potential false training of features of the codec (i.e., mislead the coder into thinking it is starting in a quiet environment), which may lead to longerthan-normal training times once the noise starts.

Figure 9 shows the physical configuration for those conditions without a mask, and Figure 10 shows the physical configuration for those conditions with a mask. The figures show the HATS in the middle of the attenuated chamber with loudspeakers around the perimeter. For those conditions without a mask, the transmitting microphone for the communication system will be positioned 5 cm (2 in.) directly in front of the lip reference point (LRP). For those conditions with a mask, the transmitting microphone of the communication system will be positioned 2.5 cm (1 in.) directly in front of the voice transmission port on the mask.

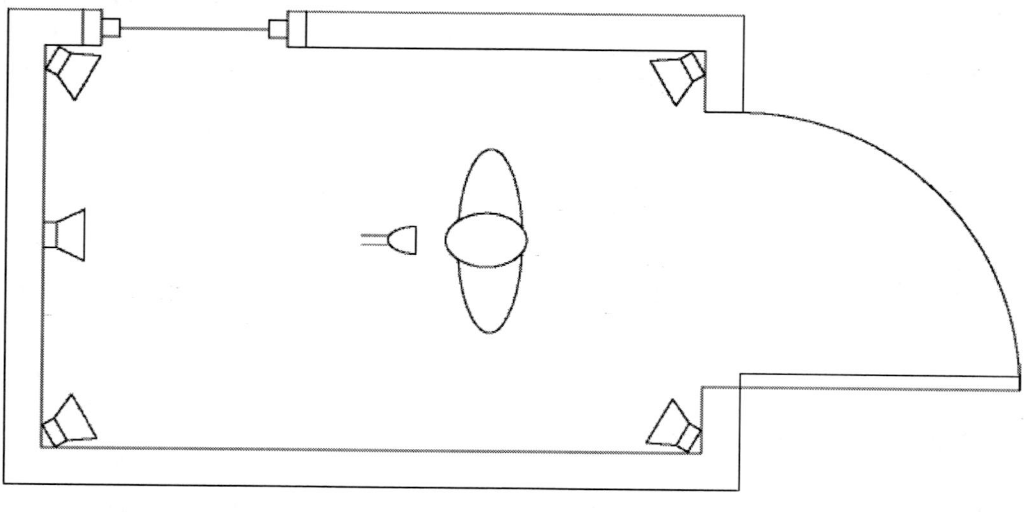

Figure 9. Physical configuration for non-mask conditions.

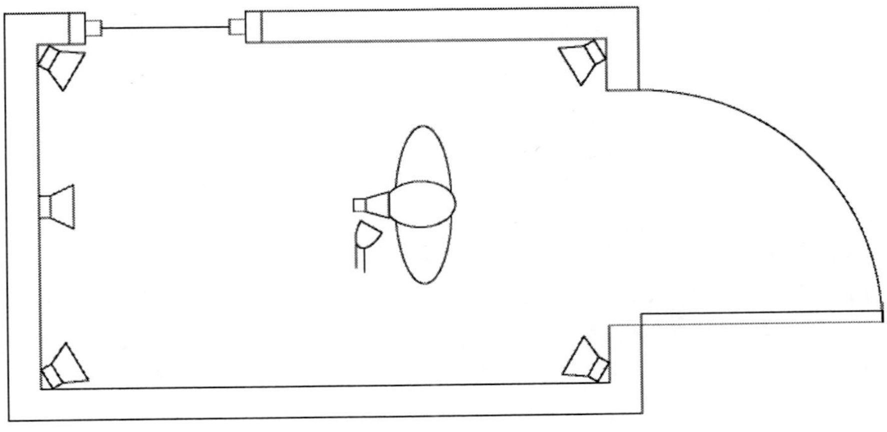

Figure 10. Physical configuration for conditions with SCBA mask.

For single point background noise sources, the sound will be generated by the loudspeakers at the front of the room (in front of the HATS). For multipoint noise sources, all five speakers will be used to generate the noise.

For the purposes of this experiment, the artificial mouth of the HATS was equalized to flat ± 1 dB in the band of 160 Hz to 10 kHz. Speech is played through the HATS at a level of 100 dBC.[5] This is consistent with measurements made on behalf of TIA of users talking in a loud noise environment.

For the receiving end, the speaker of the receiving radio will be positioned at the ear reference point (ERP). Volume on the receive radio will be set such that a 1011 Hz P25 test tone generates an acoustic signal of 85 dBA at 1" from the speaker grille. Recording will be done through the artificial ear of the second HATS in a "quiet" environment. This enables any listening environmental noise to be inserted during the actual listening experiment.

4.3. Signal and Acoustic Noise Levels

In order to construct the appropriate signal to noise levels for voice coder encoding, careful attention to the noise signal filtering and level adjustment of the signals is required. The definition for signal to noise ratio in dB for this test is given in (1).

$$SNR = ActiveSpeechLevel - 20\log_{10}(NoiseLevel) \qquad (1)$$

Active speech level is computed according to [6]. The short-term power function of each noise is assumed to be much more stationary than speech, relieving the need for an activity/threshold detector. The computation for the noise levels follows a root mean square (RMS) algorithm that is scaled to the overload point so that dB values are negative or zero and is shown in (2).

$$NoiseLevel = \frac{1}{A_{fullscale}} \sqrt{\frac{\sum_{i=1}^{N} x_i^2}{N}} \quad (2)$$

where $A_{fullscale}$ is the fullscale amplitude of the signal, N is the number of samples, and x_i is the value of the ith sample.

The noise samples are scaled by the appropriate factor to obtain the target noise level for that specific test condition and summed together to create a noise condition file for voice coder input.

5. Subjective Evaluation of Intelligibility

The subjective evaluation of intelligibility using the MRT is specified in [9] and described in [3].

Subjective testing involves the use of a number of listeners to attempt to interpret the words spoken through the communication system. The processed speech samples are obtained as described in Section 6. Because intelligibility is subject to individual abilities to interpret spoken words, reliable results are obtained when a number of listeners are used. For this assessment a total of 30 listeners are used in 6 groups of 5 listeners. To mitigate the effects of the order of presentation of the speech samples to the listeners, different presentation orders are used with each listener.

5.1. Experiment Randomization

The listening material consists of 6 lists of 50 words from each of 6 talkers (3 male, 3 female). This listening material is processed for each of the 30 test conditions yielding a total of 54,000 test samples (50 words x 6 lists x 6 talkers x 30 test conditions). Each listener will score the intelligibility of 25 words from one list for one male and one female talker for each condition.

To ensure that there are no effects of the order of presentation, the material used in the listening tests is presented to each listener in a different, randomized order.

Randomization is constrained in the following ways:

1. 25 words from one male and one female speaker for each test condition (combination of system and impairment) are presented exactly once to each listener such that no words are repeated (25x2x30 = 1500 samples per listener).
2. Listeners are effectively paired in that consecutive odd/even numbered listeners will collectively hear all 50 words from one male and one female speaker for each test condition.
3. Randomization is done in blocks of 60 test samples, such that one sample of each combination of codec, talker gender, and impairment is presented once, in each

block, to each listener. There are 5 blocks in each listening session. This means that the listeners rate each combination of codec and impairment approximately equally in the beginning, middle, and end of the session. Each listener participates in 5 sessions during the experiment.

4. The selection of the particular test samples for each block and their presentation order within a block is randomized.
5. Consecutive talkers are always different. This is accomplished by alternating the gender of the talker.

5.2. MRT Evaluation Laboratory

The Evaluation Laboratory conforms where possible to the applicable sections of ITU-T Recommendation P.800 [10]. An Evaluation Laboratory is chosen to perform the evaluations. The Evaluation Laboratory is responsible for conducting the tests as described in Section 5.3 and the delivery of the results of the experiment as described in Section 5.4.

Prior to the start of the test the listeners participate in a practice session. During this practice session they are presented with 60 practice sentences which they score. The practice sentences consist of a block of material as described in Section 7.1.1 item 3, but taken from the larger corpus of speech material, excluding the samples that they score for the experimental sessions. After the practice session the listeners are asked if they understand what they are supposed to do. If there are any questions they are answered at that time. After that the formal test begins. The purpose of the practice session is to: (a) expose the listener to the range of audio quality of the test, and (b) to see if they understand what they are supposed to be doing. This is in accordance with P.800, clause B.4.6, “Instructions to Subjects.”

The experimental results are presented, while the data from the practice sessions is discarded. The results from the listening test are then permuted to undo the randomization described in Section 5.1 so that the results can be reported as described in Section 5.4.

5.3. Intelligibility Testing

The subjective evaluation consists of one experiment to determine effects of background noise and talker variability.

The listening test for the intelligibility test follows the Modified Rhyme Test (MRT) method [3]. In this type of test, each listener listens to a sentence asking them to select a word from a prescribed list. The listeners’ ability to select the correct word is averaged across listeners and produces a percentage of intelligibility score.

The MRT consists of the test conditions shown in Table 1.

5.3.1. Conducting the Listening Test

Presentation of speech material is made via high fidelity near-field monitor speaker at a distance of 1. 5m from the listener. The delivery system is calibrated to deliver an average speech listening level of 75 – 85 dBA when measured at the listening position. The equivalent acoustic noise level of the delivery system does not exceed 35 dBA.

[Before Training Session]

Welcome and thank you for coming.

This experiment is five sessions of approximately 20 minutes each. You will be able to take a break after each session, and we will have you take at least a 5 minute break after the second session. This experiment involves no risk or discomfort, and you are free to leave the experiment at any time for any reason, it will not be a problem for us. If you have any questions about the experiment, please feel free to ask them before the experiment starts. Your responses will be kept confidential, and will only be used as part of this experiment.

This experiment uses the speakers in the room, so you will not be able to adjust the volume.

The purpose of this experiment is to gather intelligibility information on systems that might be used for communications service between separate locations. You will be hearing a number of samples of speech reproduced in the speaker. Each sample will consist of the sentence "Please select the word X" spoken by male or female speakers.

Please listen to the sentence, and then select the requested word from the list on the PDA. You may hear background noise in some of the samples. Please do your best to pick the requested word.

Any questions?

[Between Training Session and Session 1]

Any problems during the training session?

We will now do the first session of 300 samples. Any questions before we begin?

Figure 11. Instructions to test subjects. 5.4 Analysis and Reporting of Results.

Listeners are seated in a room, with an ambient pink noise level of 70 dBA as defined in Sections 8.10.4.11 through 8.10.4.15 of [9]. The exact configuration of the room and characterization of the noise is shown in Appendix E.

That the listeners are practitioners notwithstanding, the listeners should be naïve with respect to communication technology issues; that is, they are not experts in telephone design, digital voice encoding algorithms, and so on. The sample includes adults of mixed sex, age, and practitioner discipline. Persons have audiometrically normal hearing as defined in 8.10.4.3 of [9].

The test is conducted as described in Sections 8.10.5.1 and 8.10.5.2, with the exceptions that the carrier sentence will be "Please select the word [list word]" and that listeners will select the word on a touch screen.

The administration of the experiment is as follows. The processed speech is presented to a panel of 30 listeners. The 30 listeners are segmented into five listening group sessions {A B C D E F} of 6 members each. Each listening group session contains 9,000 sentences out of the possible pool of 54,000[6] sentences to evaluate. The 9,000 listening session sentences are further divided into 6 groups of 1,500 sentences for each listener of the group. Each listener will hear 5 lists of 300 sentences.

Before starting the test, the subjects are given the instructions in Figure 11. The instructions may be modified to allow for variations in laboratory data-gathering apparatus.

The results are reported in a series of tables and figures in Section 6. The analysis and reporting is outlined in Sections 8.10.5 and 8.10.6 of [9]. Averages are computed using the adjusted method recommended for closed set tests as described in Section 10.2 of [3]. An analysis of variance (ANOVA) will be computed to enable comparisons between the implementations. A more detailed description of the analysis follows.

5.4.1. Intelligibility Scores

The first result presented for each environment is a table of intelligibility scores. The intelligibility score represents the fraction of words that were selected correctly. This is the modified intelligibility score (RA) that accounts for potential correct guessing in the limited set of choices given to the listeners. This is either reported as a decimal number between 0.0 and 1.0 or as a percentage.

5.4.2. Background Information for ANOVA

ANOVA and a multiple comparison test can assist in the determination of whether there is a significant variation between the speech outputs of the three communication systems (and their respective vocoder technologies), and if so, which is better. A common multiple comparison test used in previous tests is the Tukey test.

The data under analysis with ANOVA and Tukey consists of adjusted average intelligibility scores, RA (described in Section 10.2 of [3]), collected for 3 communication systems, 6 talkers, 8 acoustic noise conditions, and 30 listeners. ANOVA compares the variance of the overall sample population with the variance within each sub-population, and if those variances exceed a value given by the Fisher F-distribution, then the null hypothesis is false. In this case, the null hypothesis is that communication systems do not make a difference in intelligibility.

The hypothesis under test, H, is that the communication system (in particular the vocoder implemented by that communication system) affects the intelligibility, RA, as measured by the MRT. The hypothesis can be tested for each of the 8 acoustic noise conditions. If the hypothesis is true, it is also desired to know which communication system is better.

For this test, the $\alpha = 0.01$ for the F-distribution in the ANOVA for the given degrees of freedom, df, and sample size, n.

5.4.3. Input to ANOVA

A table is generated for each noise condition that includes the per-talker and per-communicationsystem average and standard deviation.

Averages of means and standard deviations are computed for a sample size of 6 talkers and 30 listeners, $N_{sample} = 180$. Representative calculations of mean and standard deviation for each noise condition, for a communication system is shown in Equation 3 and Equation 4, respectively.

$$R_A(v) = \frac{\sum R_A(v,t,L)}{N_{sample}} \tag{3}$$

for vocoder v. Sum is over talkers, t=1 . .6, and listeners, L=1 -30.

$$S(v) = \sqrt{\frac{\sum\left[(R_A(v,t,L) - R_A(v))^2\right]}{N_{sample} - 1}} \tag{4}$$

From the listening test description in 5.3.1, each listener will hear 25 words (1/2 of a 50-word list) for each talker for each condition. An RA value is computed for each talker for each condition. To achieve the required sample indicated in 5.3.1, 10 different listeners will hear different groups of 25 words for each talker and condition. This results in 10 RA values for each talker and condition.

5.4.4. Example ANOVA Calculation and Multiple Comparison Test

The analysis of data will be done in the statistical analysis package Minitab™. A sample Minitab report of the computation of the ANOVA and the Tukey test is shown below. Annotations are provided after each section of the report.

One-way ANOVA: RA versus System

Source	DF	SS	MS	F	P
System	3	0.31772	0.10591	13.40	0.000
Error	236	1.86493	0.00790		
Total	239	2.18266			

S = 0.08889 R-Sq = 14.56% R-Sq(adj) = 13.47%

The above represents the ANOVA results. The value of P is compared with the statistical significance of α=0.01. If $P < \alpha$, statistically significant differences exist.

```
                                    Individual 99% CIs For Mean Based on
                                    Pooled StDev
Level   N     Mean    StDev   -------+---------+---------+---------+--
1      60  0.88080  0.07830                      (------*-------)
2      60  0.82560  0.08924              (------*-------)
3      60  0.80000  0.11184   (------*------)
4      60  0.88560  0.07075                       (------*-------)
                              -------+---------+---------+---------+--
                                 0.800     0.840     0.880     0.920

Pooled StDev = 0.08889
```

The above shows a rough graphic of the mean values and standard deviations for each of the four systems (Levels) in the test. The definition of each system is not important for this example. A pooled standard deviation value is also presented.

```
Tukey 99% Simultaneous Confidence Intervals
All Pairwise Comparisons among Levels of System

Individual confidence level = 99.81%
```

Corresponding to the α=0.01, the Tukey simultaneous confidence interval is set to 99%. This provides an individual comparison confidence interval of 99.81%.

```
System = 1 subtracted from:

System      Lower     Center      Upper
2        -0.10627   -0.05520   -0.00413
3        -0.13187   -0.08080   -0.02973
4        -0.04627    0.00480    0.05587

System      +---------+---------+---------+---------

2              (------*------)
3           (------*-------)
4                    (-------*------)
           +---------+---------+---------+---------
        -0.140    -0.070     0.000     0.070
```

This first comparison is System 1 to the other systems. If the other systems' values are all either above or below zero, then there is a statistically significant difference. If the other systems upper and lower values cross zero, then no statistically significant difference was detected. In this case, System 2 and System 3 are statistically significantly different from System 1. System 4 is not significantly different from System 1; this can also be stated as System 1 and System 4 are statistically similar.

```
System = 2 subtracted from:

System      Lower     Center      Upper
3        -0.07667   -0.02560    0.02547
4         0.00893    0.06000    0.11107

System      +---------+---------+---------+---------
3                    (-------*-------)
4                              (-------*-------)
            +---------+---------+---------+---------
         -0.140    -0.070     0.000     0.070
```

Table 2. Example Presentation of Tukey Results.

Tukey Multiple Comparison Results				
"YES" means significant difference				
System	4	3	2	1
1	NO	YES	YES	
2	YES	NO		
3	YES			
4				

The second comparison is System 2 to System 3 and System 4. In this case System 2 and System 3 are statistically similar, but System 2 and System 4 are statistically significantly different. System 1 is not included in this graph because it was previously compared to System 2.

```
System = 3 subtracted from:

System     Lower    Center     Upper      +---------+---------+---------+---------
4        0.03453   0.08560   0.13667                                (------*-------)
                                          +---------+---------+---------+---------
                                       -0.140    -0.070     0.000     0.070
```

The third comparison is System 3 to System 4. In this case, System 3 and System 4 are statistically significantly different. System 1 and System 2 are not included in this graph because they were compared with System 3 in previous sections of this Minitab report.

Once the comparisons are made, the information can be synthesized into a more readable format as demonstrated in Table 2. A tabular result is presented in Section 6 for each environment in which the ANOVA detected a significant difference. Full Minitab reports for each environment are included in Appendix A.

6. Disclosure of Test Results

This section describes the results of the experiment. One-way ANOVA results and Tukey comparisons were computed using the Minitab statistical analysis package. Input to the ANOVA consisted of 10 adjusted intelligibility scores (RA) per talker/noise environment/system. Each of those adjusted intelligibility scores were based on the intelligibility of 25 samples for that particular talker/environment/system combination that were presented to that particular listener.

The RA listener scores for the experiment are tabulated in Appendix B. The analysis of the scores for the experiment is given in Section 6.1.

6.1. Experiment Results Detail

Experimental results are presented below for each noise environment in the test. For each noise environment, a table of the adjusted intelligibility scores (RA) and an analysis of the one-way ANOVA ($p=0.01$) is presented. If a statistically significant difference was detected by the ANOVA, a Tukey comparison analysis ($\alpha = 0.01$) follows to reveal which elements are different, from a statistical standpoint.

Finally, the results for the noise environment are presented in bar charts with statistically similar systems linked together by a horizontal red bar. If the red bar includes a dotted line, the systems crossed by the dotted portion of the line are not included in the statistical similarity. For example, in Figure 12, the top red bar indicates that the 25 kHz Analog FM and the 12.5 kHz Analog FM are statistically similar and the bottom horizontal red bar is used to indicate that the P25 Full Rate and P25 Enh Full Rate are statistically similar.

Table 3. Intelligibility Scores, ANOVA Results, and Tukey Results for Environment 1.

Intelligibility Scores (RA)

25 kHz ANA FM	P25 Full Rate	P25 Enh Full Rate	12.5 kHz ANA FM
0.881	0.826	0.800	0.886

One-way ANOVA: RA versus Condition.

Source	DF	SS	MS	F	P	Significant?
Condition	3	0.31772	0.10591	13.4	0.000	YES
Error	236	1.86493	0.0079			
Total	239	2.18266				

S = 0.08889 R.Sq = 14.56% R.Sq(adj) = 13.47%

Tukey Multiple Comparison Results.

"YES" means significant difference				
System	4	3	2	1
1.25 kHzANAFM	NO	YES	YES	
2 .P25 Full Rate	YES	NO		
3 .P25 Enh Full Rate	YES			
4 .12.5 kHz ANA FM				

6.1.1. Results for Noise Environment 1 – "Clean"

The "clean" noise transmission environment consisted of the ideal communications case with no background noise and no SCBA mask. The listening environment was as described in 5.3.1. Table 3 contains the intelligibility scores, ANOVA, and Tukey results for this environment. The detailed Minitab report is found in Appendix A.

Figure 12 contains the bar chart of results for this environment. The solid portions of the top horizontal red bar in Figure 12 are used to indicate that the two analog systems are statistically similar in this environment. The bottom horizontal red bar is used to indicate that the two digital systems are statistically similar. Both of the analog systems are statistically better than the digital systems.

6.1.2. Results for Noise Environment 2 – Fire Truck Pump Panel

The fire truck pump panel noise transmission environment consisted of the communications case with 4 dB SNR and no SCBA mask. The listening environment was as described in 5.3.1. Table 4 contains the intelligibility scores, ANOVA, and Tukey results for this environment. The detailed Minitab report is found in Appendix A.

Figure 13 contains the bar chart of results for this environment. The solid portions of the horizontal red bar in Figure 13 are used to indicate that the 25 kHz Analog FM and the P25 Enhanced Full Rate are statistically equivalent in this environment. Both of these are statistically better than the P25 Full Rate.

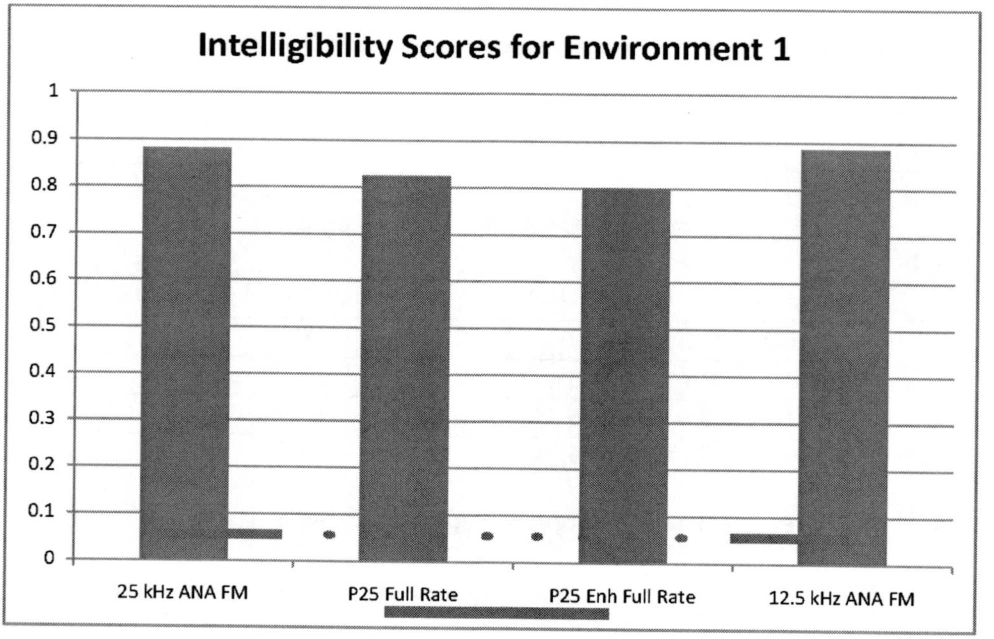

Figure 12. Intelligibility scores and statistical equivalences for environment 1.

Table 4. Intelligibility Scores, ANOVA Results, and Tukey Results for Environment 2.

Intelligibility Scores (RA)

25 kHz ANA FM	P25 Full Rate	P25 Enh Full Rate
0.437	0.341	0.470

One-way ANOVA: RA versus Condition.

Source	DF	SS	MS	F	P	Significant?
System	2	0.5376	0.2688	11.07	0.000	YES
Error	177	4.3	0.0243			
Total	179	4.8376				

S = 0.1559 R- Sq = 11.11% R- Sq(adj) = 10.11%

Tukey Multiple Comparison Results.

"YES" means significant difference				
System		3	2	1
1	25 kHz ANA FM	NO	YES	
2	P25 Full Rate	YES		
3	P25 Enh Full Rate			

6.1.3. Results for Noise Environment 3 – SCBA Mask

This noise transmission environment consists of the communications case with a user wearing an SCBA mask with no additional background noise. The listening environment was as described in 5.3.1. Table 5 contains the intelligibility scores, ANOVA, and Tukey results for this environment. The detailed Minitab report is found in Appendix A.

Figure 14 contains the bar chart of results for this environment. The solid portions of the top horizontal red bar in Figure 14 are used to indicate that the two analog systems are statistically equivalent in this environment. The bottom horizontal red bar is used to indicate that the two digital systems are statistically equivalent. Both of the analog systems are statistically better than the digital systems.

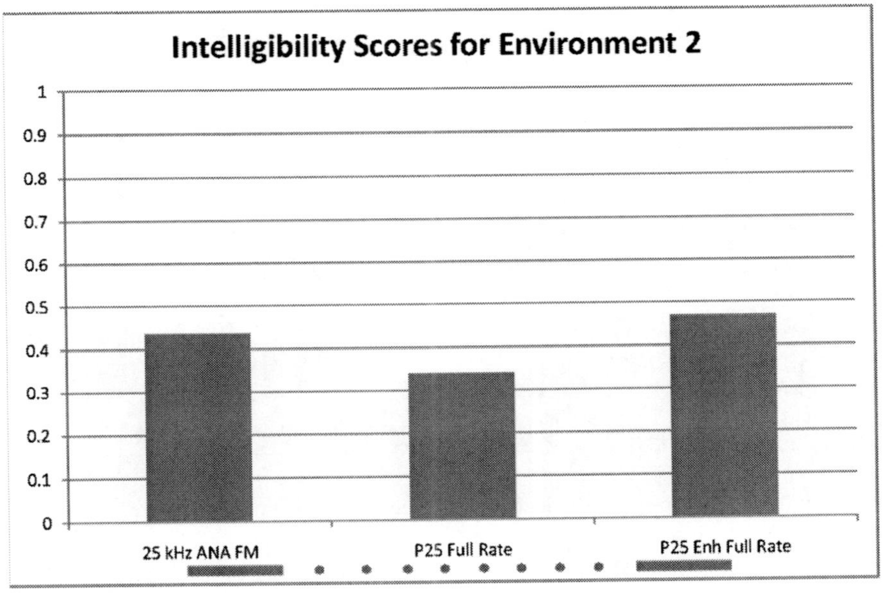

Figure 13. Intelligibility scores and statistical equivalences for environment 2.

Table 5. Intelligibility Scores, ANOVA Results, and Tukey Results for Environment 3.

Intelligibility Scores (RA)

25 kHz ANA FM	P25 Full Rate	P25 Enh Full Rate	12.5 kHz ANA FM
0.785	0.522	0.591	0.798

One- way ANOVA: RA versus Condition.

Source	DF	SS	MS	F	P	Significant?
System	3	3.47	1.1567	73.16	0.000	YES
Error	236	3.7313	0.0158			
Total	239	7.2013				

S = 0.1257 R–Sq = 48.19% R–Sq(adj) = 47.53%.

Tukey Multiple Comparison Results.

"YES"means significant difference				
System	4	3	2	1
1 – 25 kHz ANA FM	NO	YES	YES	
2 – P25 Full Rate	YES	NO		
3 – P25 Enh Full Rate	YES			
4 – 12.5 kHz ANA FM				

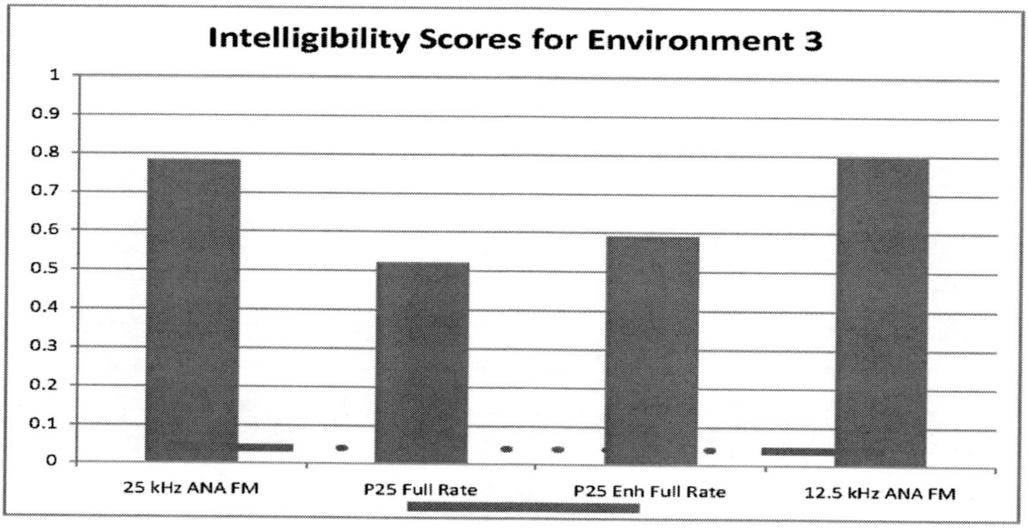

Figure 14. Intelligibility scores and statistical equivalences for environment 3.

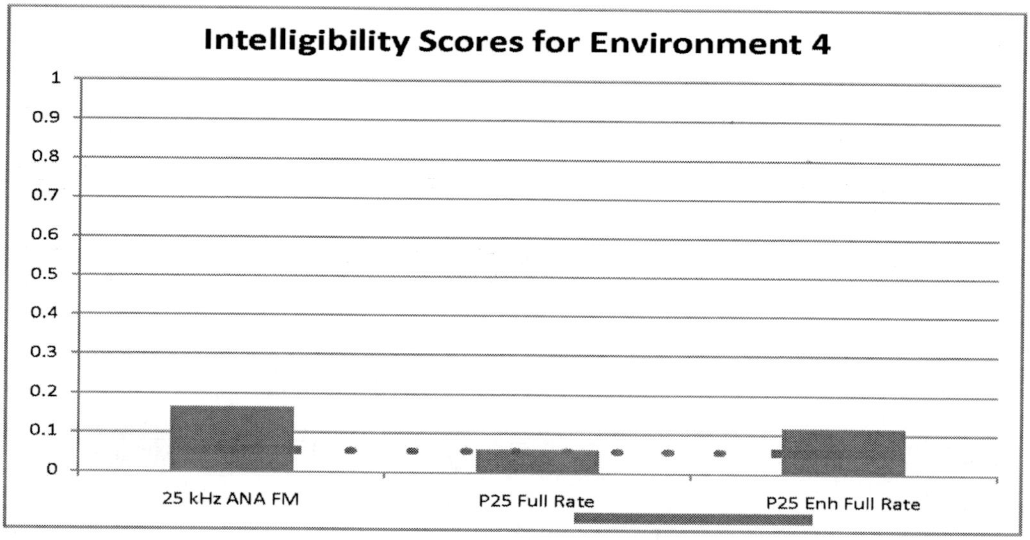

Figure 15. Intelligibility scores and statistical equivalences for environment 4.

6.1.4. Results for Noise Environment 4 – Low-Air Alarm

The low air alarm noise transmission environment consisted of the communications case where a user would be wearing a mask and need to communicate while the low-air alarm was sounding. The listening environment was as described in 5.3.1. Table 6 contains the intelligibility scores, ANOVA, and Tukey results for this environment. The detailed Minitab report is found in Appendix A.

Figure 15 contains the bar chart of results for this environment. The solid portions of the top horizontal red bar in Figure 15 are used to indicate that the 25 kHz Analog FM and P25 Enhanced Full Rate systems are statistically equivalent in this environment. The bottom horizontal red bar is used to indicate that the two digital systems are statistically equivalent in this environment. The 25 kHz Analog FM performs significantly better than the P25 Full Rate.

Table 6. Intelligibility Scores, ANOVA Results, and Tukey Results for Environment 4.

Intelligibility Scores (RA)

25 kHz ANA FM	P25 Full Rate	P25 Enh Full Rate
0.165	0.058	0.115

One- way ANOVA: RA versus Condition.

Source	DF	SS	MS	F	P	Significant?
System	2	0.3454	0.1727	11.1	0.000	Yes
Error	177	2.7534	0.0156			
Total	179	3.0988				

Tukey Multiple Comparison Results.

"YES" means significant difference			
System	3	2	1
1 – 25 kHz ANA FM	NO	YES	
2 – P25 Full Rate	NO		
3 – P25 Enh Full Rate			

6.1.5. Results for Noise Environment 5 – PASS Alarms

This noise transmission environment consisted of the communications case where a SCBA clad person would need to communicate upon finding two downed firefighters with their SCBA alarms sounding. The listening environment was as described in 5.3.1. Table 7 contains the intelligibility scores, ANOVA, and Tukey results for this environment. The detailed Minitab report is found in Appendix A.

Figure 16 contains the bar chart of results for this environment. The horizontal red bar in Figure 16 is used to indicate that the P25 Full Rate and the P25 Enhanced Full Rate are statistically equivalent in this environment. The 25 kHz Analog FM performed significantly better than either of the two digital systems.

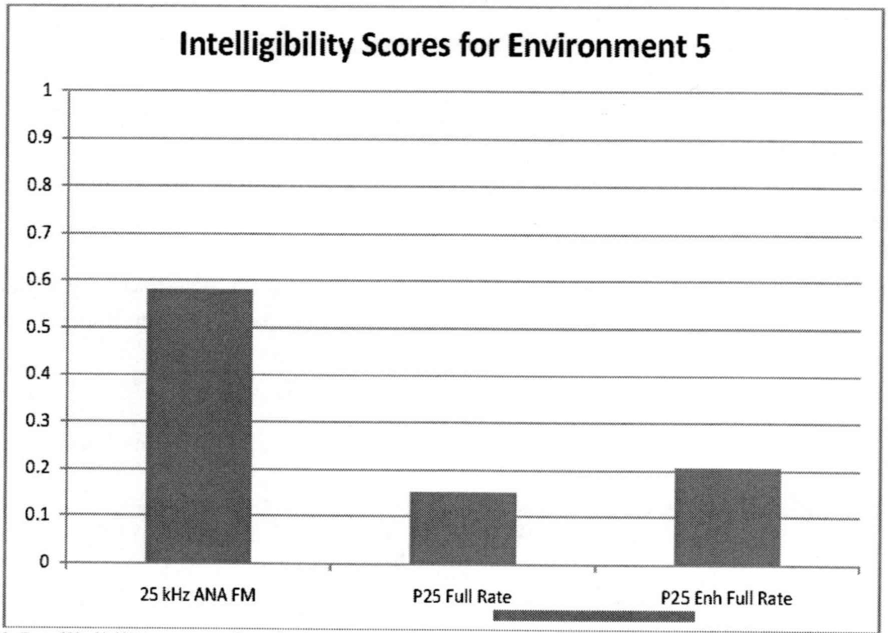

Figure 16. Intelligibility scores for environment 5.

Table 7. Intelligibility Scores, ANOVA Results, and Tukey Results for Environment 5.

Intelligibility Scores (RA)

25 kHz ANA FM	P25 Full Rate	P25 Enh Full Rate
0.581	0.152	0.206

One‒way ANOVA: RA versus Condition.

Source	DF	SS	MS	F	P	Signficant?
System	2	6.5503	3.2752	178.71	0.000	YES
Error	177	3.2438	0.0183			
Total	179	9.7942				

S = 0.1354 R- Sq = 66.88% R- Sq(adj) = 66.51%

Tukey Multiple Comparison Results.

"YES" means significant difference			
System	3	2	1
1 – 25 kHz ANA FM	YES	YES	
2 – P25 Full Rate	NO		
3 – P25 Enh Full Rate			

6.1.6. Results for Noise Environment 6 – Rotary Saw

This noise environment represents the communications case where a person wearing a mask would need to communicate in the vicinity of someone operating a gas-powered rotary saw. The listening environment was as described in 5.3.1. Table 8 contains the intelligibility scores, ANOVA, and Tukey results for this environment. The detailed Minitab report is found in Appendix A.

Figure 17 contains the bar chart of results for this environment. The horizontal red bar in Figure 17 is used to indicate that the four systems perform in a statistically equivalent manner in this environment.

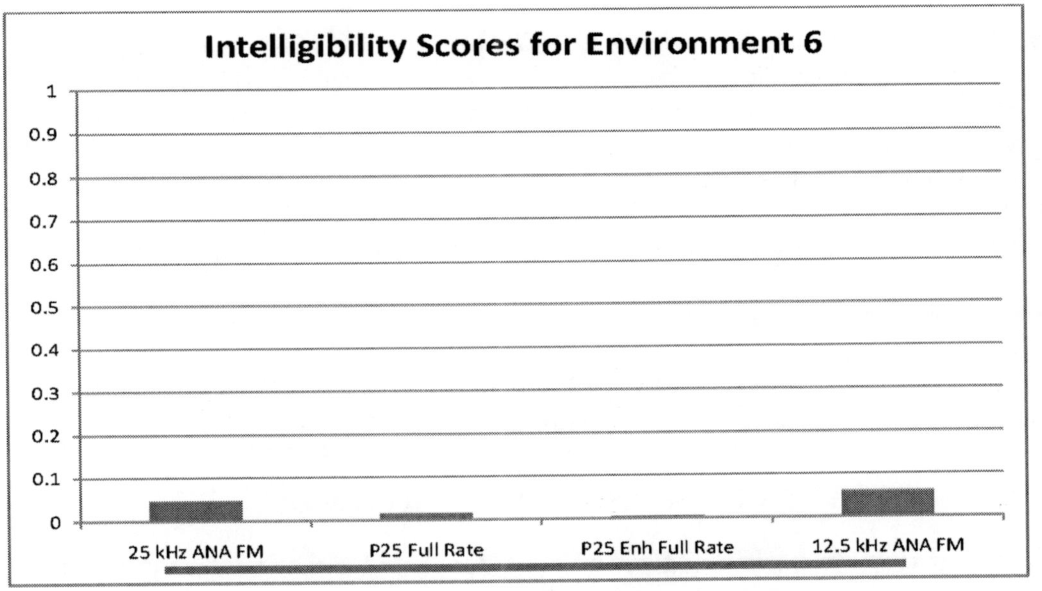

Figure 17. Intelligibility scores and statistical equivalences for environment 6.

Table 8. Intelligibility Scores, ANOVA Results, and Tukey Results for Environment 6.

Intelligibility Scores (RA)

25 kHz ANA FM	P25 Full Rate	P25 Enh Full Rate	12.5 kHz ANA FM
0.046	0.015	0.005	0.059

One-way ANOVA: RA versus Condition.

Source	DF	SS	MS	F	P	Significant?
System	3	0.11807	0.03936	3.95	0.009	YES
Error	236	2.3502	0.00996			
Total	239	2.46827				

S = 0.09979 R- Sq = 4.78% R- Sq(adj) = 3.57% .

Tukey Multiple Comparison Results.

"YES" means significant difference				
System	4	3	2	1
1 – 25 kHz ANA FM	NO	NO	NO	
2 – P25 Full Rate	NO	NO		
3 – P25 Enh Full Rate	NO			
4 – 12.5 kHz ANA FM				

6.1.7. Results for Noise Environment 7 – Chainsaw

This noise transmission environment represents the communications case where a person wearing a mask needs to communicate near someone operating a chainsaw. The listening environment was as described in 5.3.1. Table 9 contains the intelligibility scores, ANOVA, and Tukey results for this environment. The detailed Minitab report is found in Appendix A.

Table 9. Intelligibility Scores, ANOVA Results, and Tukey Results for Environment 7.

Intelligibility Scores (RA)

25 kHz ANA FM	P25 Full Rate	P25 Enh Full Rate
0.064	0.019	0.002

One-way ANOVA: RA versus Condition.

Source	DF	SS	MS	F	P	Significant?
System	2	0.1349	0.0674	5.99	0.003	YES
Error	177	1.9922	0.0113			
Total	179	2.1271				

S = 0.1061 R- Sq = 6.34% R- Sq(adj) = 5.28%

Tukey Multiple Comparison Results.

"YES" means significant difference			
System	3	2	1
1 – 25 kHz ANA FM	YES	NO	
2 – P25 Full Rate	NO		
3 – P25 Enh Full Rate			

Figure 18 contains the bar chart of results for this environment. The top horizontal red bar in Figure 18 is used to indicate that the 25 kHz Analog FM system and the P25 Full Rate system are statistically equivalent. The bottom horizontal red bar is used to indicate that the P25 Full Rate and P25 Enhanced Full Rate systems are statistically equivalent in this environment. The 25 kHz Analog FM performs significantly better than the P25 Enh Full Rate.

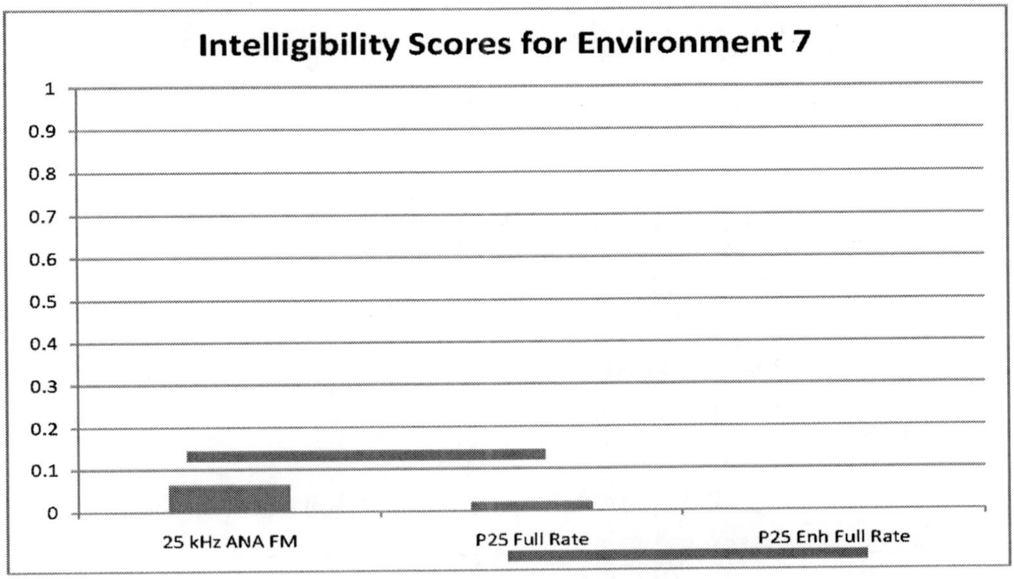

Figure 18. Intelligibility scores and statistical equivalences for environment 7.

6.1.8. Results for Noise Environment 8 – Fog Nozzle

This noise transmission environment represents the communications case where someone wearing a mask would need to communicate near a fire hose operating with a fog nozzle. The listening environment was as described in 5.3.1. Table 10 contains the intelligibility scores and ANOVA results for this environment. The detailed Minitab report is found in Appendix A.

Figure 19 contains the bar chart of results for this environment. The horizontal red bar in Figure 19 indicates that there were no detectable differences between the systems.

Table 10. Intelligibility Scores and ANOVA Results for Environment 8.

Intelligibility Scores (RA)

25 kHz ANA FM	P25 Full Rate	P25 Enh Full Rate
0.110	0.068	0.075

One-way ANOVA: RA versus Condition.

Source	DF	SS	MS	F	P	Significant?
System	2	0.0593	0.0297	2.35	0.098	NO
Error	177	2.233	0.0126			
Total	179	2.2923				

Tukey analysis is not presented because the ANOVA did not detect a significant difference.

Table 11. Intelligibility Scores, ANOVA Results, and Tukey Results for Environment 9.

Intelligibility Scores (RA)

25 kHz ANA FM	P25 Full Rate	P25 Enh Full Rate
0.039	0.015	0.092

One-way ANOVA: RA versus Condition.

Source	DF	SS	MS	F	P	Significant?
System	2	0.1852	0.0926	7.27	0.001	YES
Error	177	2.2546	0.0127			
Total	179	2.4398				

S = 0.1129 R- Sq = 7.59% R- Sq(adj) = 6.55% .

Tukey Multiple Comparison Results.

"YES" means significant difference			
System	3	2	1
1 – 25 kHz ANA FM	NO	NO	
2 – P25 Full Rate	YES		
3 – P25 Enh Full Rate			

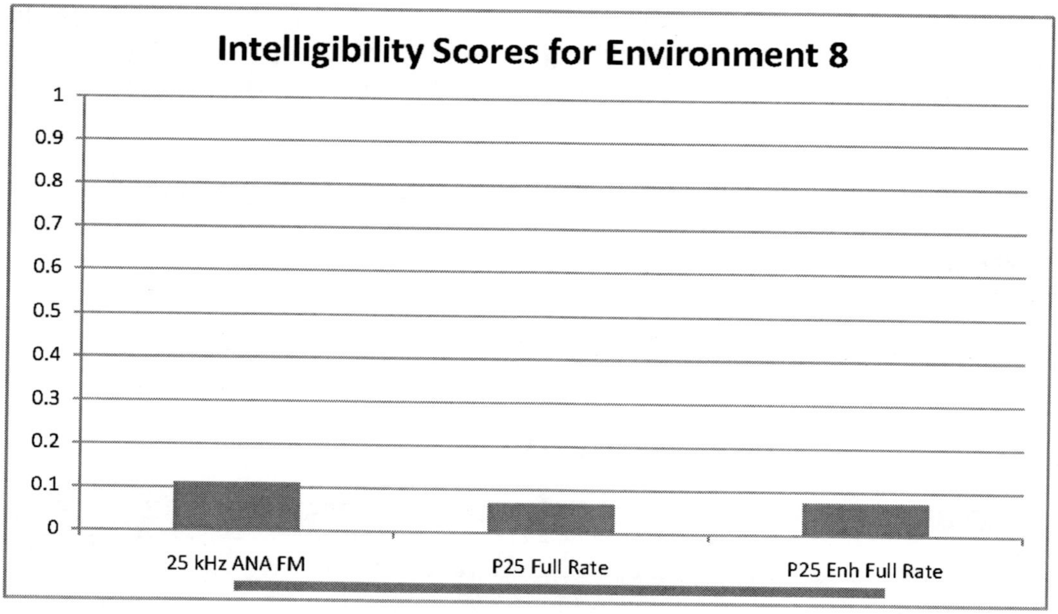

Figure 19. Intelligibility scores and statistical equivalences for environment 8.

6.1.9. Results for Noise Environment 9 – Amplified Mask

This noise transmission environment represents the communications case where a person with a voice amplifier attached to their mask would need to communicate near a person operating a rotary saw. The listening environment was as described in 5.3.1. Table 11 contains the intelligibility scores, ANOVA, and Tukey results for this environment. The detailed Minitab report is found in Appendix A.

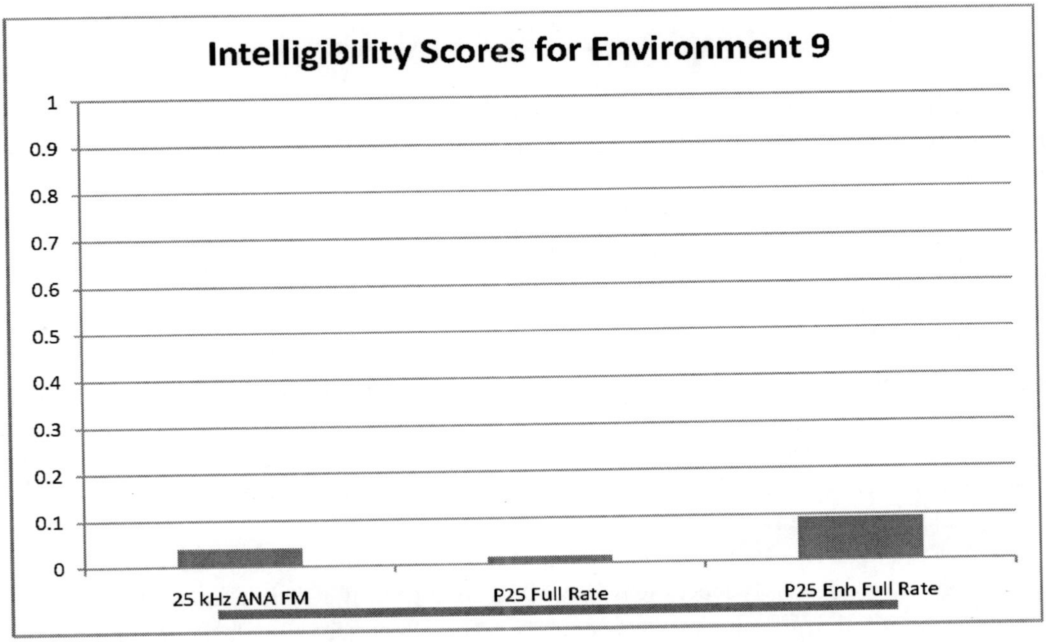

Figure 20. Intelligibility scores and statistical equivalences for environment 9.

Figure 20 contains the bar chart of results for this environment. The horizontal red bar in Figure 20 is used to indicate that the 25 kHz Analog FM is statistically equivalent to the P25 Full Rate and P25 Enhanced Full Rate systems for this environment. The P25 Enhanced Full Rate system performs significantly better than the P25 Full Rate system.

SUMMARY AND CONCLUSIONS

Based on emerging concerns among fire service practitioners, this report describes an experiment that was conducted to measure intelligibility of radio systems in the presence of fireground noise. The test method selected to evaluate intelligibility was the Modified Rhyme Test.

Four communication systems (25 kHz analog, 12.5 kHz analog, Project 25 Full Rate, and Project 25 Enhanced Full Rate) were included in the test. Nine transmission environments (listed below) were also identified for the test.

Environment 1.	No background noise, no mask (referred to as the Clean condition)
Environment 2.	Fire truck pump panel, no mask
Environment 3.	Mask with no background noise
Environment 4.	Two Personal Alert Safety System (PASS) alarms, with mask
Environment 5.	In-mask low air alarm
Environment 6.	Rotary saw cutting metal garage door, with mask
Environment 7.	Chainsaw cutting wood, with mask
Environment 8.	21/2" hose with fog nozzle, with mask
Environment 9.	Rotary saw cutting metal garage door, with amplified mask

The report describes the method for generating the recordings of noisy speech data, the requirements of the listening environment, and the process for presenting the recorded speech to listeners.

The results of the subjective testing show that there are environments where analog radios had higher intelligibility than the digital radios, and also that there are environments where none of the systems perform very well. Observations included the following:

- The MRT test seems to be significantly more sensitive (i.e., more able to discriminate) to differences in communication paths than MOS tests. This is evidenced by the much greater number of occurrences of statistically significant differences between the P25 Full Rate and P25 Enhanced Full Rate vocoder systems and between the analog and digital systems than has occurred in previous MOS-based tests such as [12] and [13].
- Without masks or background noise (i.e., Environment 1) all four communication systems effectively meet the NFPA 1981-2007 goal of 80% intelligibility.
- Four of the nine noise environments were effectively too difficult for all tested communication systems, with all systems well below 10%. In future testing, either noise reduction technologies may be employed to evaluate these environments, or the environments may be adapted to provide more useful results. Another alternative might be context-based intelligibility testing.
- In environment 3 (SCBA mask with no additional background noise), the two analog systems effectively maintain the NFPA 198 1-2007 goal of 80%[7] intelligibility through a mask, while the digital systems were 52% for the P25 Full Rate vocoder system and 59% for the P25 Enhanced Full Rate vocoder system. This implies that there may need to be some additional considerations when using these digital communication systems in environments where a communicator must wear a SCBA mask for safety purposes.
- In environment 2 (fire truck pump panel noise without an SCBA mask), both the analog system and the P25 Enhanced Full Rate system vocoder performed significantly better than the P25 Full Rate system.
- The largest performance difference between the analog and digital systems was environment 5 (SCBA mask with two PASS alarms).
- Analog FM outperformed P25 Full Rate 5 of the 9 environments tested. In the remaining four environments, all radios had equivalent performance in three of them.

In one environment, the performance of the P25 Full Rate system was equivalent to that of Analog FM.

- Analog FM outperformed P25 Enhanced Full rate in 4 of the 9 environments tested. In the remaining 5 environments, all radios had equivalent performance in three of them. In two environments, the performance of the P25 Enhanced Full Rate system was equivalent to that of Analog FM.
- In the case of P25 Full Rate and P25 Enhanced Full Rate, the two were statistically equivalent in 8 of 9 environments, with the P25 Enhanced Full Rate system performing better than the P25 Full Rate in 1 of the 9 environments.
- In the three environments where 12.5 kHz Analog FM was included, it performed equivalent with 25 kHz Analog FM. This system should be evaluated further.
- The testing indicated that there are significant communication challenges regarding SCBA and other fire safety equipment. In the case of a mask with no additional background noise at the transmission site, there was a notable degradation in intelligibility for the digital systems. In addition, some low air alarm designs and PASS alarms may interfere with communications intelligibility.

The information contained in this report may be of value to those planning to purchase and deploy new radio communication systems in their agencies. The information here should NOT be the only factor utilized in the purchasing process. Other factors should include agencies involved, current assets, departmental operating procedures and policies, budgeting plans, spectrum availability, State Communications Interoperability Plans (SCIPs), and specific situations faced by the purchasing agency that are not covered in this report.

REFERENCES

Project 25 Vocoder Description, TIA-102.BABA.

Project 25 Mean Opinion Score Conformance Test, TIA-102.BABB.

ANSI S3.2, *American National Standard Method for Measuring the Intelligibility of Speech Over Communication Systems*, 1989.

A.V. Oppenheim, R.W. Schafer, *Digital Signal Processing*, Englewood Cliffs, NJ: Prentice Hall Inc., 1975.

F.J. Harris, "On the use of Windows for Harmonic Analysis with the Discrete Fourier Transform," *Proceedings of the IEEE*, Vol. 66, No. 1, Jan. 1978.

ITU-T Recommendation P.56, "*Objective measurement of active speech level.*"

ITU-T Recommendation P.57, "*Artificial Ears.*"

ITU-T Recommendation P.58, "*Head and Torso Simulator (HATS) for Telephonometry.*"

NFPA 1981 *Standard on Open-Circuit Self-Contained Breathing Apparatus (SCBA) for Emergency Services,* 2007 Edition.

ITU-T Recommendation P.800, "*Methods for subjective determination of transmission quality.*"

ITU-T Recommendation G. 191, "*Software Tools for Speech and Audio Coding Standardization.*"

Project 25 Vocoder Evaluation Mean Opinion Score Test, TIA TSB 102.BABE.

Experiment 3 MOS Test Plan for Vocoder Technology for Project 25, TIA TSB 1 02.BABF.

APPENDIX A: CONDITION LABELS AND MINITAB REPORTS

This appendix contains the representation of specific condition number to a communication system and background noise. Table A-1 contains the list of condition numbers for this experiment.

Table A-1. Condition Number, Communication System, and Noise Environment.

Condition Label	Communication System	Background Noise
1	25 kHz Analog FM	No mask, no background noise
2	25 kHz Analog FM	No mask, fire truck pump panel (4 dB S/N)
3	25 kHz Analog FM	Mask, no background noise
4	25 kHz Analog FM	Mask, in-mask low air alarm (15 dB S/N)
5	25 kHz Analog FM	Mask, two PASS alarms (-2 dB S/N)
6	25 kHz Analog FM	Mask, rotary saw (4 dB S/N)
7	25 kHz Analog FM	Mask, chainsaw (5 dB S/N)
8	25 kHz Analog FM	Mask, 21/2" hose with fog nozzle (9 dB S/N)
9	25 kHz Analog FM	Amplified mask, rotary saw (4 dB S/N)
10	P25 Full Rate	No mask, no background noise
11	P25 Full Rate	No mask, fire truck pump panel (4 dB S/N)
12	P25 Full Rate	Mask, no background noise
13	P25 Full Rate	Mask, in-mask low air alarm (15 dB S/N)
14	P25 Full Rate	Mask, two PASS alarms (-2 dB S/N)
15	P25 Full Rate	Mask, rotary saw (4 dB S/N)
16	P25 Full Rate	Mask, chainsaw (5 dB S/N)
17	P25 Full Rate	Mask, 21/2" hose with fog nozzle (9 dB S/N)
18	P25 Full Rate	Amplified mask, rotary saw (4 dB S/N)
19	P25 Enh Full Rate	No mask, no background noise
20	P25 Enh Full Rate	No mask, fire truck pump panel (4 dB S/N)
21	P25 Enh Full Rate	Mask, no background noise
22	P25 Enh Full Rate	Mask, in-mask low air alarm (15 dB S/N)
23	P25 Enh Full Rate	Mask, two PASS alarms (-2 dB S/N)
24	P25 Enh Full Rate	Mask, rotary saw (4 dB S/N)
25	P25 Enh Full Rate	Mask, chainsaw (5 dB S/N)
26	P25 Enh Full Rate	Mask, 21/2" hose with fog nozzle (9 dB S/N)
27	P25 Enh Full Rate	Amplified mask, rotary saw (4 dB S/N)
28	12.5 kHz Analog FM	No mask, no background noise
29	12.5 kHz Analog FM	Mask, no background noise
30	12.5 kHz Analog FM	Mask, rotary saw (4 dB S/N)

Sections A. 1 through A.9 contain the full Minitab reports for Environments 1 through 9, respectively. For help interpreting the Minitab reports, see Section 5.4.4.

A.1. Minitab Report for Environment 1

One-way ANOVA: RA versus System

```
Source   DF       SS       MS      F      P
System    3  0.31772  0.10591  13.40  0.000
Error   236  1.86493  0.00790
Total   239  2.18266

S = 0.08889   R-Sq = 14.56%   R-Sq(adj) = 13.47%


                                 Individual 99% CIs For Mean Based on
                                 Pooled StDev
Level   N     Mean    StDev   -------+---------+---------+---------+--
1      60  0.88080  0.07830                        (------*-------)
2      60  0.82560  0.08924             (------*-------)
3      60  0.80000  0.11184   (------*------)
4      60  0.88560  0.07075                          (------*-------)
                              -------+---------+---------+---------+--
                                   0.800     0.840     0.880     0.920

Pooled StDev = 0.08889


Tukey 99% Simultaneous Confidence Intervals
All Pairwise Comparisons among Levels of System

Individual confidence level = 99.81%


System = 1 subtracted from:

System     Lower    Center     Upper
2       0.10627   0.05520   0.00413
3      -0.13187  -0.08080  -0.02973
4      -0.04627   0.00480   0.05587

System      +---------+------------------+---------
2                (------*------)
3            (------*-------)
4                       (-------*------)
            +---------+------------------+---------
         -0.140    -0.070     0.000     0.070


System = 2 subtracted from:

System     Lower    Center     Upper
3      -0.07667  -0.02560   0.02547
4       0.00893   0.06000   0.11107

System      +---------+------------------+---------
3                    (------*-------)
4                               (-------*------)
            +---------+------------------+---------
         0.140     0.070      0.000     0.070


System = 3 subtracted from:

System     Lower    Center     Upper     +---------+---------+------------------
4        0.03453   0.08560   0.13667                              (------*-------)
                                         +---------+---------+------------------
                                      -0.140    -0.070     0.000     0.070
```

A.2. Minitab Report for Environment 2

One-way ANOVA: RA versus System

```
Source   DF      SS      MS      F      P
System    2  0.5376  0.2688  11.07  0.000
Error   177  4.3000  0.0243
Total   179  4.8376

S = 0.1559   R-Sq = 11.11%   R-Sq(adj) = 10.11%


                               Individual 99% CIs For Mean Based on
                               Pooled StDev
Level   N    Mean   StDev  --+---------+---------+---------+-------
1      60  0.4368  0.1383                 (--------*--------)
2      60  0.3408  0.1815  (--------*--------)
3      60  0.4696  0.1443                       (-------*--------)
                           --+---------+---------+---------+-------
                           0.300     0.360     0.420     0.480

Pooled StDev = 0.1559


Tukey 99% Simultaneous Confidence Intervals
All Pairwise Comparisons among Levels of System

Individual confidence level = 99.64%


System = 1 subtracted from:

System    Lower   Center    Upper  --------+---------+---------+---------+-
2       -0.1799  -0.0960  -0.0121     (------*------)
3       -0.0511   0.0328   0.1167            (------*------)
                                   --------+---------+---------+---------+-
                                       -0.12      0.00      0.12      0.24


System = 2 subtracted from:

System   Lower  Center   Upper  --------+---------+---------+---------+-
3       0.0449  0.1288  0.2127                     (------*------)
                                --------+---------+---------+---------+-
                                    -0.12      0.00      0.12      0.24
```

A.3. Minitab Report for Environment 3

One-way ANOVA: RA versus System

```
Source   DF      SS      MS      F      P
System    3  3.4700  1.1567  73.16  0.000
Error   236  3.7313  0.0158
Total   239  7.2013

S = 0.1257   R-Sq = 48.19%   R-Sq(adj) = 47.53%

                                  Individual 99% CIs For Mean Based on
                                  Pooled StDev
Level   N    Mean   StDev  --+---------+---------+---------+-------
1      60  0.7848  0.1010                              (---*----)
2      60  0.5216  0.1600  (---*---)
3      60  0.5912  0.1353         (---*---)
4      60  0.7984  0.0955                                (---*---)
                           --+---------+---------+---------+-------
                           0.50      0.60      0.70      0.80

Pooled StDev = 0.1257


Tukey 99% Simultaneous Confidence Intervals
All Pairwise Comparisons among Levels of System

Individual confidence level = 99.81%


System = 1 subtracted from:

System     Lower   Center    Upper  -------+---------+---------+---------+--
2        -0.3354  -0.2632  -0.1910  (---*--)
3        -0.2658  -0.1936  -0.1214      (--*---)
4        -0.0586   0.0136   0.0858            (---*--)
                                    -------+---------+---------+---------+--
                                       -0.20      0.00      0.20      0.40


System = 2 subtracted from:

System     Lower  Center   Upper  -------+---------+---------+---------+--
3        -0.0026  0.0696  0.1418                 (--*---)
4         0.2046  0.2768  0.3490                           (---*--)
                                  -------+---------+---------+---------+--
                                     -0.20      0.00      0.20      0.40


System = 3 subtracted from:

System    Lower  Center   Upper  -------+---------+---------+---------+--
4        0.1350  0.2072  0.2794                           (--*---)
                                 -------+---------+---------+---------+--
                                    -0.20      0.00      0.20      0.40
```

A.4. Minitab Report for Environment 4

One-way ANOVA: RA versus System

```
Source   DF      SS      MS      F      P
System    2  0.3454  0.1727  11.10  0.000
Error   177  2.7534  0.0156
Total   179  3.0988

S = 0.1247   R-Sq = 11.15%   R-Sq(adj) = 10.14%


                               Individual 99% CIs For Mean Based on
                               Pooled StDev
Level   N    Mean   StDev   -------+---------+---------+---------+--
1      60  0.1648  0.1450                    (-------*-------)
2      60  0.0576  0.1086   (--------*-------)
3      60  0.1152  0.1177             (-------*-------)
                            -------+---------+---------+---------+--
                                 0.050     0.100     0.150     0.200

Pooled StDev = 0.1247


Tukey 99% Simultaneous Confidence Intervals
All Pairwise Comparisons among Levels of System

Individual confidence level = 99.64%


System = 1 subtracted from:

System    Lower   Center    Upper   --+---------+---------+---------+------
2       -0.1743  -0.1072  -0.0401   (--------*-------)
3       -0.1167  -0.0496   0.0175         (--------*-------)
                                    --+---------+---------+---------+------
                                   -0.160    -0.080     0.000     0.080


System = 2 subtracted from:

System    Lower  Center   Upper   --+---------+---------+---------+-------
3       -0.0095  0.0576  0.1247                        (-------*--------)
                                  --+---------+---------+---------+-------
                                 -0.160    -0.080     0.000     0.080
```

A.5. Minitab Report for Environment 5

One-way ANOVA: RA versus System

```
Source   DF      SS      MS       F      P
System    2  6.5503  3.2752  178.71  0.000
Error   177  3.2438  0.0183
Total   179  9.7942

S = 0.1354   R-Sq = 66.88%   R-Sq(adj) = 66.51%


                             Individual 99% CIs For Mean Based on
                             Pooled StDev
Level   N    Mean   StDev    ---+---------+---------+---------+------
1      60  0.5808  0.1259                                (--*--)
2      60  0.1520  0.1255    (--*--)
3      60  0.2056  0.1529        (--*--)
                             ---+---------+---------+---------+------
                              0.15      0.30      0.45      0.60

Pooled StDev = 0.1354


Tukey 99% Simultaneous Confidence Intervals
All Pairwise Comparisons among Levels of System

Individual confidence level = 99.64%


System = 1 subtracted from:

System    Lower   Center    Upper     -+---------+---------+---------+--------
2       -0.5017  -0.4288  -0.3559     (---*----)
3       -0.4481  -0.3752  -0.3023        (----*---)
                                      -+---------+---------+---------+--------
                                    -0.48     -0.32     -0.16      0.00


System = 2 subtracted from:

System    Lower  Center   Upper     -+---------+---------+---------+--------
3       -0.0193  0.0536  0.1265                                (---*----)
                                    -+---------+---------+---------+--------
                                  -0.48     -0.32     -0.16      0.00
```

A.6. Minitab Report for Environment 6

One-way ANOVA: RA versus System

```
Source   DF       SS       MS     F      P
System    3  0.11807  0.03936  3.95  0.009
Error   236  2.35020  0.00996
Total   239  2.46827

S = 0.09979   R-Sq = 4.78%   R-Sq(adj) = 3.57%

                                Individual 99% CIs For Mean Based on
                                Pooled StDev
Level   N     Mean    StDev   --------+---------+---------+---------+-
1      60  0.04640  0.11125             (---------*---------)
2      60  0.01520  0.08636      (---------*---------)
3      60  0.00480  0.09263   (---------*---------)
4      60  0.05920  0.10686                (---------*---------)
                              --------+---------+---------+---------+-
                                  0.000     0.035     0.070     0.105

Pooled StDev = 0.09979


Tukey 99% Simultaneous Confidence Intervals
All Pairwise Comparisons among Levels of System

Individual confidence level = 99.81%


System = 1 subtracted from:

System     Lower    Center    Upper  ---------+---------+---------+---------+
2       -0.08853  -0.03120  0.02613     (---------*---------)
3       -0.09893  -0.04160  0.01573    (---------*---------)
4       -0.04453   0.01280  0.07013           (---------*---------)
                                     ---------+---------+---------+---------+
                                           -0.060     0.000     0.060     0.120


System = 2 subtracted from:

System     Lower    Center    Upper  ---------+---------+---------+---------+
3       -0.06773  -0.01040  0.04693        (---------*---------)
4       -0.01333   0.04400  0.10133               (---------*---------)
                                     ---------+---------+---------+---------+
                                           -0.060     0.000     0.060     0.120


System = 3 subtracted from:

System     Lower    Center    Upper  ---------+---------+---------+---------+
4       -0.00293   0.05440  0.11173                (---------*---------)
                                     ---------+---------+---------+---------+
                                           -0.060     0.000     0.060     0.120
```

A.7. Minitab Report for Environment 7

One-way ANOVA: RA versus System

```
Source   DF      SS      MS     F      P
System    2  0.1349  0.0674  5.99  0.003
Error   177  1.9922  0.0113
Total   179  2.1271

S = 0.1061   R-Sq = 6.34%   R-Sq(adj) = 5.28%

                                Individual 99% CIs For Mean Based on Pooled StDev
Level   N     Mean   StDev     -+---------+---------+---------+--------
1      60   0.0640  0.1181                          (---------*---------)
2      60   0.0192  0.0990                (---------*----------)
3      60  -0.0016  0.1002      (----------*---------)
                               -+---------+---------+---------+--------
                            -0.035     0.000     0.035     0.070

Pooled StDev = 0.1061


Tukey 99% Simultaneous Confidence Intervals
All Pairwise Comparisons among Levels of System

Individual confidence level = 99.64%


System = 1 subtracted from:

System    Lower   Center    Upper    +---------+---------+---------+---------
2       -0.1019  -0.0448   0.0123       (---------*--------)
3       -0.1227  -0.0656  -0.0085    (--------*---------)
                                     +---------+---------+---------+---------
                                  -0.120    -0.060     0.000     0.060


System = 2 subtracted from:

System    Lower   Center    Upper    +---------+---------+---------+---------
3       -0.0779  -0.0208   0.0363              (---------*--------)
                                     +---------+---------+---------+---------
                                  -0.120    -0.060     0.000     0.060
```

A.8. Minitab Report for Environment 8

One-way ANOVA: RA versus System

```
Source   DF      SS      MS     F      P
System    2  0.0593  0.0297  2.35  0.098
Error   177  2.2330  0.0126
Total   179  2.2923

S = 0.1123   R-Sq = 2.59%   R-Sq(adj) = 1.49%


                             Individual 99% CIs For Mean Based on Pooled StDev
Level   N    Mean   StDev    +---------+---------+---------+---------
1      60  0.1096  0.1187                   (-----------*-----------)
2      60  0.0680  0.1020    (-----------*-----------)
3      60  0.0752  0.1155      (-----------*-----------)
                             +---------+---------+---------+---------
                           0.030     0.060     0.090     0.120


Pooled StDev = 0.1123
```

A.9 Minitab Report for Environment 9

One-way ANOVA: RA versus System

```
Source   DF      SS      MS     F      P
System    2  0.1852  0.0926  7.27  0.001
Error   177  2.2546  0.0127
Total   179  2.4398

S = 0.1129   R-Sq = 7.59%   R-Sq(adj) = 6.55%


                                Individual 99% CIs For Mean Based on
                                Pooled StDev
Level   N    Mean   StDev  ------+---------+---------+---------+---
1      60  0.0392  0.1113        (---------*--------)
2      60  0.0152  0.0768  (---------*--------)
3      60  0.0920  0.1412                      (--------*--------)
                           ------+---------+---------+---------+---
                               0.000     0.040     0.080     0.120

Pooled StDev = 0.1129


Tukey 99% Simultaneous Confidence Intervals
All Pairwise Comparisons among Levels of System

Individual confidence level = 99.64%


System = 1 subtracted from:

System     Lower   Center   Upper    +---------+---------+---------+---------
2        -0.0848  -0.0240  0.0368           (--------*-------)
3        -0.0080   0.0528  0.1136                     (--------*-------)
                                     +---------+---------+---------+---------
                                  -0.140    -0.070     0.000     0.070


System = 2 subtracted from:

System   Lower  Center   Upper    +---------+---------+---------+---------
3       0.0160  0.0768  0.1376                        (--------*--------)
                                  +---------+---------+---------+---------
                               -0.140    -0.070     0.000     0.070
```

Appendix B: Listener Scores

This appendix tabulates the scores collected from the listener panels. It contains the listener number, the talker number, the condition number, and the adjusted intelligibility score (RA).

Table B-1. Scores Collected from Listener Panels.

Listener	Talker	Condition	RA
1	F1	4	0.28
1	F1	12	0.616
1	F1	15	0.04
1	F1	16	–0.008
1	F1	18	0.088
1	F1	19	0.856
1	F1	22	0.232
1	F1	28	0.856
1	F3	1	0.904
1	F3	2	0.52
1	F3	5	0.664
1	F3	6	–0.008
1	F3	7	0.088
1	F3	8	0.136
1	F3	10	0.808
1	F3	13	0.088
1	F3	14	0.328
1	F3	23	0.088
1	F3	25	0.088
1	F3	26	0.136
1	F3	27	0.328
1	F4	3	0.76
1	F4	9	–0.104
1	F4	11	0.568
1	F4	17	0.04
1	F4	20	0.664
1	F4	21	0.568
1	F4	24	0.04
1	F4	29	0.904
1	F4	30	0.136
1	M1	4	–0.104
1	M1	12	0.568
1	M1	15	0.136
1	M1	16	0.088
1	M1	18	0.088

Table B-1. (Continued).

Listener	Talker	Condition	RA
1	M1	19	0.76
1	M1	22	0.088
1	M1	28	0.856
1	M3	1	0.712
1	M3	2	0.088
1	M3	5	0.568
1	M3	6	–0.008
1	M3	7	0.04
1	M3	8	0.04
1	M3	10	–0.808
1	M3	13	0.104
1	M3	14	0.04
1	M3	23	0.008
1	M3	25	0.04
1	M3	26	0.008
1	M3	27	0.184
1	M4	3	0.76
1	M4	9	–0.008
1	M4	11	0.232
1	M4	17	–0.008
1	M4	20	0.52
1	M4	21	0.76
1	M4	24	–0.008
1	M4	29	0.808
1	M4	30	–0.056
2	F1	4	0.616
2	F1	12	0.808
2	F1	15	–0.056
2	F1	16	0.136
2	F1	18	0.136
2	F1	19	1
2	F1	22	0.232
2	F1	28	0.952
2	F3	1	0.856
2	F3	2	0.568
2	F3	5	0.76
2	F3	6	0.184
2	F3	7	0.184
2	F3	8	0.04
2	F3	10	0.76
2	F3	13	0.088
2	F3	14	0.28

Listener	Talker	Condition	RA
2	F3	23	0.376
2	F3	25	0.088
2	F3	26	0.04
2	F3	27	0.472
2	F4	3	0.856
2	F4	9	0.136
2	F4	11	0.616
2	F4	17	0.088
2	F4	20	0.616
2	F4	21	0.664
2	F4	24	–0.056
2	F4	29	0.808
2	F4	30	0.088
2	M1	4	0.136
2	M1	12	0.568
2	M1	15	0.184
2	M1	16	0.088
2	M1	18	0.04
2	M1	19	1
2	M1	22	0.28
2	M1	28	0.904
2	M3	1	0.952
2	M3	2	0.184
2	M3	5	0.712
2	M3	6	0.04
2	M3	7	0.04
2	M3	8	0.088
2	M3	10	0.856
2	M3	13	0.088
2	M3	14	0.088
2	M3	23	0.04
2	M3	25	0.04
2	M3	26	0.136
2	M3	27	–0.008
2	M4	3	0.952
2	M4	9	–0.008
2	M4	11	0.472
2	M4	17	0.088
2	M4	20	0.232
2	M4	21	0.568
2	M4	24	0.088
2	M4	29	0.952

Table B-1. (Continued).

Listener	Talker	Condition	RA
2	M4	30	0.04
3	F1	3	0.712
3	F1	8	0.04
3	F1	9	0.184
3	F1	10	0.856
3	F1	17	–0.056
3	F1	21	0.76
3	F1	25	– 0.056
3	F1	26	–0.008
3	F1	27	0.184
3	F1	30	0.232
3	F3	11	0.136
3	F3	12	0.424
3	F3	20	0.616
3	F3	24	–0.104
3	F3	28	0.712
3	F3	29	0.856
3	F4	1	0.856
3	F4	2	0.472
3	F4	4	–0.008
3	F4	5	0.472
3	F4	6	0.184
3	F4	7	0.04
3	F4	13	0.136
3	F4	14	0.28
3	F4	15	0.088
3	F4	16	0.088
3	F4	18	–0.152
3	F4	19	0.904
3	F4	22	0.184
3	F4	23	0.136
3	M1	3	0.616
3	M1	8	0.136
3	M1	9	–0.008
3	M1	10	0.904
3	M1	17	–0.056
3	M1	21	0.568
3	M1	25	0.088
3	M1	26	0.04
3	M1	27	0.088
3	M1	30	–0.104
3	M3	11	0.04

Listener	Talker	Condition	RA
3	M3	12	0.424
3	M3	20	0.424
3	M3	24	0.04
3	M3	28	0.904
3	M3	29	0.664
3	M4	1	0.952
3	M4	2	0.424
3	M4	4	0.088
3	M4	5	0.424
3	M4	6	–0.008
3	M4	7	–0.056
3	M4	13	–0.056
3	M4	14	–0.104
3	M4	15	–0.008
3	M4	16	–0.104
3	M4	18	–0.008
3	M4	19	0.76
3	M4	22	–0.008
3	M4	23	–0.056
4	F1	3	0.808
4	F1	8	0.328
4	F1	9	0.328
4	F1	10	0.904
4	F1	17	0.088
4	F1	21	0.712
4	F1	25	–0.056
4	F1	26	0.376
4	F1	27	0.52
4	F1	30	0.232
4	F3	11	0.472
4	F3	12	0.568
4	F3	20	0.568
4	F3	24	–0.008
4	F3	28	0.904
4	F3	29	0.856
4	F4	1	0.904
4	F4	2	0.568
4	F4	4	0.136
4	F4	5	0.424
4	F4	6	–0.008
4	F4	7	0.088
4	F4	13	0.232

Table B-1. (Continued).

Listener	Talker	Condition	RA
4	F4	14	0.088
4	F4	15	–0.008
4	F4	16	–0.056
4	F4	18	–0.008
4	F4	19	0.808
4	F4	22	–0.104
4	F4	23	0.088
4	M1	3	0.808
4	M1	8	0.136
4	M1	9	0.184
4	M1	10	0.904
4	M1	17	–0.056
4	M1	21	0.616
4	M1	25	–0.152
4	M1	26	0.04
4	M1	27	0.136
4	M1	30	0.04
4	M3	11	0.136
4	M3	12	0.184
4	M3	20	0.424
4	M3	24	–0.008
4	M3	28	0.808
4	M3	29	0.664
4	M4	1	0.952
4	M4	2	0.52
4	M4	4	0.136
4	M4	5	0.52
4	M4	6	–0.056
4	M4	7	0.088
4	M4	13	–0.008
4	M4	14	0.232
4	M4	15	–0.104
4	M4	16	–0.008
4	M4	18	–0.008
4	M4	19	0.952
4	M4	22	0.184
4	M4	23	0.184
5	F1	1	0.952
5	F1	2	0.424
5	F1	5	0.424
5	F1	6	–0.008
5	F1	7	0.088

Listener	Talker	Condition	RA
5	F1	11	0.76
5	F1	13	0.328
5	F1	14	0.184
5	F1	20	0.568
5	F1	23	0.472
5	F1	24	–0.056
5	F1	29	0.856
5	F3	3	0.808
5	F3	4	0.232
5	F3	9	–0.152
5	F3	15	–0.056
5	F3	16	0.04
5	F3	17	–0.056
5	F3	18	0.088
5	F3	19	0.808
5	F3	21	0.616
5	F3	22	0.088
5	F3	30	0.04
5	F4	8	0.184
5	F4	10	0.856
5	F4	12	0.52
5	F4	25	0.088
5	F4	26	0.088
5	F4	27	–0.008
5	F4	28	0.808
5	M1	1	0.952
5	M1	2	0.328
5	M1	5	0.52
5	M1	6	–0.056
5	M1	7	–0.056
5	M1	11	0.136
5	M1	13	0.088
5	M1	14	0.136
5	M1	20	0.328
5	M1	23	0.136
5	M1	24	–0.056
5	M1	29	0.712
5	M3	3	0.664
5	M3	4	–0.056
5	M3	9	–0.056
5	M3	15	0.04
5	M3	16	0.28

Table B-1. (Continued).

Listener	Talker	Condition	RA
5	M3	17	0.088
5	M3	18	–0.056
5	M3	19	0.76
5	M3	21	0.28
5	M3	22	0.232
5	M3	30	–0.056
5	M4	8	–0.008
5	M4	10	0.664
5	M4	12	0.328
5	M4	25	0.136
5	M4	26	0.328
5	M4	27	0.184
5	M4	28	0.952
6	F1	1	0.856
6	F1	2	0.28
6	F1	5	0.76
6	F1	6	0.088
6	F1	7	0.088
6	F1	11	0.52
6	F1	13	0.04
6	F1	14	0.568
6	F1	20	0.616
6	F1	23	0.52
6	F1	24	0.04
6	F1	29	0.808
6	F3	3	0.664
6	F3	4	0.136
6	F3	9	0.088
6	F3	15	–0.056
6	F3	16	–0.056
6	F3	17	0.136
6	F3	18	–0.008
6	F3	19	0.904
6	F3	21	0.472
6	F3	22	0.28
6	F3	30	0.136
6	F4	8	0.04
6	F4	10	0.856
6	F4	12	0.52
6	F4	25	–0.056
6	F4	26	0.088
6	F4	27	–0.056

Listener	Talker	Condition	RA
6	F4	28	0.904
6	M1	1	0.856
6	M1	2	0.616
6	M1	5	0.712
6	M1	6	–0.104
6	M1	7	0.04
6	M1	11	0.232
6	M1	13	0.04
6	M1	14	0.232
6	M1	20	0.52
6	M1	23	0.328
6	M1	24	0.088
6	M1	29	0.856
6	M3	3	0.856
6	M3	4	0.04
6	M3	9	–0.008
6	M3	15	0.04
6	M3	16	0.04
6	M3	17	0.328
6	M3	18	–0.008
6	M3	19	0.76
6	M3	21	0.568
6	M3	22	–0.104
6	M3	30	0.04
6	M4	8	0.136
6	M4	10	0.952
6	M4	12	0.76
6	M4	25	–0.008
6	M4	26	–0.008
6	M4	27	–0.008
6	M4	28	0.856
7	F1	2	0.328
7	F1	4	0.184
7	F1	5	0.856
7	F1	6	0.28
7	F1	7	0.136
7	F1	8	0.184
7	F1	9	0.328
7	F1	10	0.856
7	F1	11	0.424
7	F1	13	–0.008
7	F1	17	0.136

Table B-1. (Continued).

Listener	Talker	Condition	RA
7	F1	19	0.712
7	F1	20	0.616
7	F1	27	0.136
7	F3	1	0.712
7	F3	12	0.712
7	F3	14	0.136
7	F3	16	–0.008
7	F3	18	–0.056
7	F3	21	0.568
7	F3	23	0.28
7	F3	24	0.04
7	F3	25	–0.104
7	F4	3	0.712
7	F4	15	0.04
7	F4	22	0.136
7	F4	26	0.136
7	F4	28	0.856
7	F4	29	0.856
7	F4	30	0.04
7	M1	2	0.424
7	M1	4	0.04
7	M1	5	0.616
7	M1	6	0.184
7	M1	7	0.04
7	M1	8	–0.056
7	M1	9	–0.008
7	M1	10	0.856
7	M1	11	0.328
7	M1	13	0.04
7	M1	17	0.04
7	M1	19	0.76
7	M1	20	0.424
7	M1	27	–0.056
7	M3	1	0.856
7	M3	12	0.328
7	M3	14	0.04
7	M3	16	0.088
7	M3	18	–0.008
7	M3	21	0.328
7	M3	23	–0.008
7	M3	24	–0.104
7	M3	25	0.04

Listener	Talker	Condition	RA
7	M4	3	0.76
7	M4	15	–0.056
7	M4	22	–0.056
7	M4	26	–0.008
7	M4	28	0.856
7	M4	29	0.856
7	M4	30	0.088
8	F1	2	0.52
8	F1	4	0.376
8	F1	5	0.808
8	F1	6	0.232
8	F1	7	0.184
8	F1	8	0.28
8	F1	9	0.232
8	F1	10	0.808
8	F1	11	0.616
8	F1	13	0.136
8	F1	17	–0.056
8	F1	19	0.712
8	F1	20	0.568
8	F1	27	0.136
8	F3	1	0.856
8	F3	12	0.52
8	F3	14	0.088
8	F3	16	–0.104
8	F3	18	–0.056
8	F3	21	0.568
8	F3	23	0.232
8	F3	24	–0.056
8	F3	25	0.088
8	F4	3	0.856
8	F4	15	0.088
8	F4	22	0.136
8	F4	26	0.184
8	F4	28	0.952
8	F4	29	0.952
8	F4	30	0.088
8	M1	2	0.424
8	M1	4	0.184
8	M1	5	0.616
8	M1	6	0.088
8	M1	7	0.04

Table B-1. (Continued).

Listener	Talker	Condition	RA
8	M1	8	0.136
8	M1	9	–0.104
8	M1	10	0.904
8	M1	11	0.184
8	M1	13	0.088
8	M1	17	–0.008
8	M1	19	0.904
8	M1	20	0.76
8	M1	27	–0.2
8	M3	1	0.904
8	M3	12	0.28
8	M3	14	–0.008
8	M3	16	0.136
8	M3	18	–0.056
8	M3	21	0.568
8	M3	23	0.184
8	M3	24	–0.056
8	M3	25	–0.104
8	M4	3	0.856
8	M4	15	–0.152
8	M4	22	0.136
8	M4	26	0.136
8	M4	28	0.904
8	M4	29	0.904
8	M4	30	–0.008
9	F1	15	0.04
9	F1	16	0.088
9	F1	18	–0.008
9	F1	25	0.184
9	F1	29	0.952
9	F3	2	0.472
9	F3	3	0.856
9	F3	4	0.088
9	F3	6	0.184
9	F3	8	0.088
9	F3	9	–0.056
9	F3	10	0.904
9	F3	11	0.472
9	F3	17	–0.008
9	F3	22	0.136
9	F3	26	0.04
9	F3	27	0.136

Listener	Talker	Condition	RA
9	F3	28	1
9	F3	30	0.136
9	F4	1	1
9	F4	5	0.616
9	F4	7	0.04
9	F4	12	0.568
9	F4	13	–0.008
9	F4	14	0.136
9	F4	19	0.808
9	F4	20	0.616
9	F4	21	0.664
9	F4	23	0.088
9	F4	24	0.136
9	M1	15	–0.008
9	M1	16	0.136
9	M1	18	0.04
9	M1	25	0.088
9	M1	29	0.76
9	M3	2	0.328
9	M3	3	0.856
9	M3	4	0.136
9	M3	6	0.136
9	M3	8	0.28
9	M3	9	–0.008
9	M3	10	0.856
9	M3	11	0.232
9	M3	17	0.04
9	M3	22	–0.104
9	M3	26	–0.008
9	M3	27	–0.056
9	M3	28	0.904
9	M3	30	0.088
9	M4	1	0.904
9	M4	5	0.568
9	M4	7	0.184
9	M4	12	0.52
9	M4	13	0.04
9	M4	14	0.088
9	M4	19	0.856
9	M4	20	0.376
9	M4	21	0.712
9	M4	23	0.136

Table B-1. (Continued).

Listener	Talker	Condition	RA
9	M4	24	–0.008
10	F1	15	–0.056
10	F1	16	–0.056
10	F1	18	0.136
10	F1	25	0.088
10	F1	29	0.76
10	F3	2	0.376
10	F3	3	0.664
10	F3	4	0.184
10	F3	6	–0.008
10	F3	8	–0.008
10	F3	9	–0.056
10	F3	10	0.856
10	F3	11	0.328
10	F3	17	–0.008
10	F3	22	0.184
10	F3	26	–0.056
10	F3	27	0.184
10	F3	28	0.952
10	F3	30	–0.056
10	F4	1	0.856
10	F4	5	0.664
10	F4	7	0.136
10	F4	12	0.664
10	F4	13	0.184
10	F4	14	0.088
10	F4	19	0.808
10	F4	20	0.52
10	F4	21	0.568
10	F4	23	0.184
10	F4	24	0.04
10	M1	15	–0.008
10	M1	16	0.04
10	M1	18	0.04
10	M1	25	–0.008
10	M1	29	0.712
10	M3	2	0.184
10	M3	3	0.808
10	M3	4	–0.056
10	M3	6	–0.152
10	M3	8	0.088
10	M3	9	0.088

Listener	Talker	Condition	RA
10	M3	10	0.808
10	M3	11	–0.008
10	M3	17	0.136
10	M3	22	0.04
10	M3	26	0.04
10	M3	27	–0.104
10	M3	28	0.76
10	M3	30	–0.056
10	M4	1	0.856
10	M4	5	0.616
10	M4	7	–0.056
10	M4	12	0.568
10	M4	13	–0.104
10	M4	14	0.088
10	M4	19	0.904
10	M4	20	0.52
10	M4	21	0.808
10	M4	23	0.04
10	M4	24	0.184
11	F1	1	0.76
11	F1	3	0.952
11	F1	12	0.808
11	F1	14	0.328
11	F1	21	0.568
11	F1	22	0.232
11	F1	23	0.376
11	F1	24	0.136
11	F1	26	0.04
11	F1	28	0.952
11	F1	30	0.328
11	F3	5	0.712
11	F3	7	0.136
11	F3	13	–0.008
11	F3	15	–0.008
11	F3	19	0.712
11	F3	20	0.472
11	F3	29	0.808
11	F4	2	0.52
11	F4	4	0.136
11	F4	6	–0.104
11	F4	8	0.184
11	F4	9	0.184

Table B-1. (Continued).

Listener	Talker	Condition	RA
11	F4	10	0.904
11	F4	11	0.616
11	F4	16	0.088
11	F4	17	–0.056
11	F4	18	0.184
11	F4	25	–0.008
11	F4	27	0.088
11	M1	1	0.904
11	M1	3	0.76
11	M1	12	0.424
11	M1	14	0.136
11	M1	21	0.376
11	M1	22	0.136
11	M1	23	0.136
11	M1	24	–0.152
11	M1	26	0.04
11	M1	28	0.808
11	M1	30	0.04
11	M3	5	0.616
11	M3	7	–0.104
11	M3	13	0.184
11	M3	15	–0.008
11	M3	19	0.856
11	M3	20	0.28
11	M3	29	0.664
11	M4	2	0.424
11	M4	4	0.232
11	M4	6	0.184
11	M4	8	0.088
11	M4	9	–0.056
11	M4	10	0.712
11	M4	11	0.328
11	M4	16	0.04
11	M4	17	0.184
11	M4	18	0.184
11	M4	25	0.088
11	M4	27	0.088
12	F1	1	0.904
12	F1	3	0.904
12	F1	12	0.568
12	F1	14	0.328
12	F1	21	0.616

Listener	Talker	Condition	RA
12	F1	22	0.136
12	F1	23	0.28
12	F1	24	0.04
12	F1	26	0.136
12	F1	28	0.76
12	F1	30	0.088
12	F3	5	0.712
12	F3	7	0.04
12	F3	13	–0.104
12	F3	15	–0.104
12	F3	19	0.712
12	F3	20	0.712
12	F3	29	0.76
12	F4	2	0.184
12	F4	4	0.088
12	F4	6	0.184
12	F4	8	0.136
12	F4	9	0.136
12	F4	10	0.856
12	F4	11	0.376
12	F4	16	–0.008
12	F4	17	0.04
12	F4	18	0.088
12	F4	25	0.232
12	F4	27	0.088
12	M1	1	1
12	M1	3	0.616
12	M1	12	0.616
12	M1	14	0.136
12	M1	21	0.76
12	M1	22	–0.104
12	M1	23	0.376
12	M1	24	–0.008
12	M1	26	–0.104
12	M1	28	0.904
12	M1	30	0.04
12	M3	5	0.568
12	M3	7	–0.152
12	M3	13	–0.152
12	M3	15	–0.056
12	M3	19	0.76
12	M3	20	0.328

Table B-1. (Continued).

Listener	Talker	Condition	RA
12	M3	29	0.76
12	M4	2	0.52
12	M4	4	0.088
12	M4	6	–0.008
12	M4	8	0.136
12	M4	9	0.28
12	M4	10	0.856
12	M4	11	0.52
12	M4	16	0.232
12	M4	17	0.136
12	M4	18	0.04
12	M4	25	0.136
12	M4	27	–0.056
13	F1	1	1
13	F1	4	0.376
13	F1	11	0.616
13	F1	18	–0.008
13	F1	21	0.808
13	F1	22	0.28
13	F1	23	0.328
13	F1	25	0.088
13	F1	28	1
13	F3	2	0.376
13	F3	3	0.904
13	F3	7	0.136
13	F3	8	0.28
13	F3	14	0.232
13	F3	15	–0.056
13	F3	17	0.04
13	F3	26	0.04
13	F3	29	0.856
13	F3	30	0.04
13	F4	5	0.664
13	F4	6	0.088
13	F4	9	0.04
13	F4	10	0.904
13	F4	12	0.712
13	F4	13	0.184
13	F4	16	–0.104
13	F4	19	0.952
13	F4	20	0.616
13	F4	24	–0.104

Listener	Talker	Condition	RA
13	F4	27	–0.056
13	M1	1	1
13	M1	4	0.136
13	M1	11	0.184
13	M1	18	–0.008
13	M1	21	0.568
13	M1	22	0.376
13	M1	23	0.04
13	M1	25	–0.056
13	M1	28	0.856
13	M3	2	0.424
13	M3	3	0.712
13	M3	7	–0.008
13	M3	8	–0.056
13	M3	14	–0.056
13	M3	15	0.136
13	M3	17	0.28
13	M3	26	–0.104
13	M3	29	0.808
13	M3	30	–0.008
13	M4	5	0.376
13	M4	6	–0.104
13	M4	9	–0.104
13	M4	10	0.952
13	M4	12	0.664
13	M4	13	0.088
13	M4	16	0.04
13	M4	19	0.76
13	M4	20	0.328
13	M4	24	–0.008
13	M4	27	0.088
14	F1	1	0.856
14	F1	4	0.472
14	F1	11	0.616
14	F1	18	0.088
14	F1	21	0.616
14	F1	22	0.184
14	F1	23	0.424
14	F1	25	0.088
14	F1	28	0.952
14	F3	2	0.328
14	F3	3	0.808

Table B-1. (Continued).

Listener	Talker	Condition	RA
14	F3	7	0.04
14	F3	8	–0.056
14	F3	14	0.28
14	F3	15	–0.2
14	F3	17	0.136
14	F3	26	0.136
14	F3	29	0.76
14	F3	30	0.136
14	F4	5	0.712
14	F4	6	–0.008
14	F4	9	0.088
14	F4	10	0.856
14	F4	12	0.472
14	F4	13	0.232
14	F4	16	–0.056
14	F4	19	0.808
14	F4	20	0.424
14	F4	24	–0.056
14	F4	27	0.088
14	M1	1	0.904
14	M1	4	0.28
14	M1	11	0.28
14	M1	18	0.088
14	M1	21	0.616
14	M1	22	–0.008
14	M1	23	0.328
14	M1	25	–0.008
14	M1	28	0.904
14	M3	2	0.472
14	M3	3	0.856
14	M3	7	–0.152
14	M3	8	0.088
14	M3	14	–0.008
14	M3	15	0.088
14	M3	17	0.136
14	M3	26	0.04
14	M3	29	0.52
14	M3	30	0.136
14	M4	5	0.472
14	M4	6	–0.056
14	M4	9	–0.056
14	M4	10	0.904

Listener	Talker	Condition	RA
14	M4	12	0.472
14	M4	13	0.088
14	M4	16	–0.056
14	M4	19	0.904
14	M4	20	0.328
14	M4	24	0.04
14	M4	27	0.328
15	F1	5	0.52
15	F1	7	0.136
15	F1	12	0.664
15	F1	13	0.184
15	F1	15	0.184
15	F1	19	0.952
15	F1	20	0.472
15	F1	24	0.04
15	F1	26	0.328
15	F1	27	0.232
15	F1	29	0.712
15	F1	30	0.28
15	F3	4	0.328
15	F3	6	0.088
15	F3	9	0.136
15	F3	10	0.856
15	F3	11	0.328
15	F3	16	0.04
15	F3	18	–0.104
15	F3	25	–0.152
15	F3	28	0.952
15	F4	1	1
15	F4	2	0.664
15	F4	3	0.712
15	F4	8	0.232
15	F4	14	0.232
15	F4	17	0.136
15	F4	21	0.424
15	F4	22	0.184
15	F4	23	0.088
15	M1	5	0.424
15	M1	7	–0.104
15	M1	12	0.568
15	M1	13	0.136
15	M1	15	–0.008

Table B-1. (Continued).

Listener	Talker	Condition	RA
15	M1	19	0.712
15	M1	20	0.616
15	M1	24	–0.008
15	M1	26	–0.008
15	M1	27	0.184
15	M1	29	0.904
15	M1	30	0.088
15	M3	4	–0.104
15	M3	6	–0.152
15	M3	9	–0.008
15	M3	10	0.856
15	M3	11	–0.008
15	M3	16	–0.056
15	M3	18	–0.056
15	M3	25	–0.104
15	M3	28	0.952
15	M4	1	0.712
15	M4	2	0.472
15	M4	3	0.76
15	M4	8	0.232
15	M4	14	0.184
15	M4	17	0.04
15	M4	21	0.568
15	M4	22	0.136
15	M4	23	0.28
16	F1	5	0.664
16	F1	7	0.088
16	F1	12	0.76
16	F1	13	0.136
16	F1	15	–0.104
16	F1	19	1
16	F1	20	0.616
16	F1	24	–0.056
16	F1	26	–0.056
16	F1	27	0.28
16	F1	29	0.808
16	F1	30	0.232
16	F3	4	0.088
16	F3	6	0.184
16	F3	9	–0.152
16	F3	10	0.76
16	F3	11	0.232

Listener	Talker	Condition	RA
16	F3	16	–0.056
16	F3	18	0.04
16	F3	25	–0.008
16	F3	28	0.856
16	F4	1	0.952
16	F4	2	0.52
16	F4	3	0.664
16	F4	8	0.184
16	F4	14	0.088
16	F4	17	–0.056
16	F4	21	0.424
16	F4	22	0.28
16	F4	23	0.136
16	M1	5	0.472
16	M1	7	–0.008
16	M1	12	0.424
16	M1	13	0.136
16	M1	15	0.088
16	M1	19	0.856
16	M1	20	0.424
16	M1	24	0.04
16	M1	26	–0.008
16	M1	27	0.088
16	M1	29	0.76
16	M1	30	–0.008
16	M3	4	0.04
16	M3	6	–0.008
16	M3	9	0.04
16	M3	10	0.856
16	M3	11	0.184
16	M3	16	–0.008
16	M3	18	0.04
16	M3	25	–0.104
16	M3	28	0.808
16	M4	1	0.904
16	M4	2	0.472
16	M4	3	0.808
16	M4	8	0.088
16	M4	14	0.088
16	M4	17	0.04
16	M4	21	0.712
16	M4	22	0.088

Table B-1. (Continued).

Listener	Talker	Condition	RA
16	M4	23	0.136
17	F1	2	0.424
17	F1	3	0.712
17	F1	6	0.088
17	F1	8	–0.008
17	F1	9	0.04
17	F1	10	0.856
17	F1	14	0.184
17	F1	16	–0.008
17	F1	17	0.04
17	F3	1	0.808
17	F3	5	0.52
17	F3	12	0.28
17	F3	13	0.088
17	F3	19	0.808
17	F3	20	0.424
17	F3	21	0.712
17	F3	22	–0.104
17	F3	23	0.28
17	F3	24	–0.008
17	F3	27	0.136
17	F4	4	0.28
17	F4	7	0.136
17	F4	11	0.52
17	F4	15	–0.008
17	F4	18	–0.056
17	F4	25	–0.056
17	F4	26	0.04
17	F4	28	0.856
17	F4	29	0.904
17	F4	30	0.088
17	M1	2	0.376
17	M1	3	0.664
17	M1	6	–0.008
17	M1	8	–0.104
17	M1	9	–0.056
17	M1	10	0.856
17	M1	14	0.184
17	M1	16	0.184
17	M1	17	0.136
17	M3	1	0.76
17	M3	5	0.472

Listener	Talker	Condition	RA
17	M3	12	0.376
17	M3	13	–0.056
17	M3	19	0.616
17	M3	20	0.28
17	M3	21	0.424
17	M3	22	0.088
17	M3	23	–0.056
17	M3	24	0.04
17	M3	27	0.04
17	M4	4	0.184
17	M4	7	0.088
17	M4	11	0.376
17	M4	15	0.136
17	M4	18	–0.056
17	M4	25	–0.152
17	M4	26	0.04
17	M4	28	0.904
17	M4	29	0.712
17	M4	30	–0.056
18	F1	2	0.568
18	F1	3	0.904
18	F1	6	0.232
18	F1	8	0.136
18	F1	9	0.088
18	F1	10	0.808
18	F1	14	0.184
18	F1	16	–0.008
18	F1	17	0.328
18	F3	1	1
18	F3	5	0.472
18	F3	12	0.808
18	F3	13	0.184
18	F3	19	0.904
18	F3	20	0.472
18	F3	21	0.712
18	F3	22	0.04
18	F3	23	0.232
18	F3	24	– 0.056
18	F3	27	0.232
18	F4	4	0.184
18	F4	7	0.184
18	F4	11	0.52

Table B-1. (Continued).

Listener	Talker	Condition	RA
18	F4	15	0.136
18	F4	18	– 0.008
18	F4	25	0.04
18	F4	26	0.28
18	F4	28	1
18	F4	29	0.952
18	F4	30	0.04
18	M1	2	0.472
18	M1	3	0.904
18	M1	6	–0.008
18	M1	8	0.04
18	M1	9	–0.008
18	M1	10	0.904
18	M1	14	0.088
18	M1	16	–0.152
18	M1	17	0.136
18	M3	1	0.904
18	M3	5	0.52
18	M3	12	0.136
18	M3	13	0.088
18	M3	19	0.712
18	M3	20	0.424
18	M3	21	0.568
18	M3	22	–0.008
18	M3	23	0.376
18	M3	24	0.088
18	M3	27	–0.008
18	M4	4	0.328
18	M4	7	–0.056
18	M4	11	0.52
18	M4	15	0.136
18	M4	18	–0.008
18	M4	25	0.088
18	M4	26	0.04
18	M4	28	0.904
18	M4	29	0.904
18	M4	30	0.04
19	F1	1	0.808
19	F1	2	0.568
19	F1	8	0.136
19	F1	14	0.328
19	F1	15	0.04

Listener	Talker	Condition	RA
19	F1	20	0.472
19	F1	22	0.232
19	F1	26	0.184
19	F1	28	0.76
19	F3	4	0.088
19	F3	6	0.136
19	F3	7	–0.152
19	F3	11	0.088
19	F3	12	0.424
19	F3	13	0.04
19	F3	17	0.088
19	F3	19	0.712
19	F3	21	0.52
19	F3	25	–0.2
19	F4	3	0.712
19	F4	5	0.808
19	F4	9	–0.008
19	F4	10	0.808
19	F4	16	–0.056
19	F4	18	–0.008
19	F4	23	0.184
19	F4	24	0.088
19	F4	27	0.136
19	F4	29	0.76
19	F4	30	–0.056
19	M1	1	0.76
19	M1	2	0.424
19	M1	8	–0.008
19	M1	14	0.04
19	M1	15	0.136
19	M1	20	0.568
19	M1	22	0.232
19	M1	26	–0.008
19	M1	28	0.856
19	M3	4	0.088
19	M3	6	–0.008
19	M3	7	–0.056
19	M3	11	0.184
19	M3	12	0.376
19	M3	13	–0.056
19	M3	17	0.232
19	M3	19	0.52

Table B-1. (Continued).

Listener	Talker	Condition	RA
19	M3	21	0.328
19	M3	25	–0.104
19	M4	3	0.712
19	M4	5	0.568
19	M4	9	0.04
19	M4	10	0.856
19	M4	16	–0.056
19	M4	18	0.088
19	M4	23	0.136
19	M4	24	–0.008
19	M4	27	–0.056
19	M4	29	0.76
19	M4	30	–0.056
20	F1	1	0.952
20	F1	2	0.472
20	F1	8	0.232
20	F1	14	0.184
20	F1	15	–0.008
20	F1	20	0.424
20	F1	22	0.184
20	F1	26	0.328
20	F1	28	0.952
20	F3	4	0.328
20	F3	6	–0.056
20	F3	7	0.136
20	F3	11	0.184
20	F3	12	0.472
20	F3	13	–0.056
20	F3	17	–0.104
20	F3	19	0.904
20	F3	21	0.568
20	F3	25	0.04
20	F4	3	0.808
20	F4	5	0.664
20	F4	9	–0.056
20	F4	10	0.76
20	F4	16	–0.008
20	F4	18	0.136
20	F4	23	0.088
20	F4	24	0.04
20	F4	27	0.088
20	F4	29	0.808

Listener	Talker	Condition	RA
20	F4	30	–0.152
20	M1	1	0.76
20	M1	2	0.52
20	M1	8	–0.056
20	M1	14	0.28
20	M1	15	–0.056
20	M1	20	0.328
20	M1	22	0.056
20	M1	26	0.088
20	M1	28	0.856
20	M3	4	0.04
20	M3	6	0.04
20	M3	7	0.04
20	M3	11	0.232
20	M3	12	0.376
20	M3	13	–0.056
20	M3	17	–0.008
20	M3	19	0.76
20	M3	21	0.472
20	M3	25	–0.056
20	M4	3	0.664
20	M4	5	0.376
20	M4	9	–0.008
20	M4	10	0.664
20	M4	16	–0.056
20	M4	18	0.04
20	M4	23	0.232
20	M4	24	–0.008
20	M4	27	0.088
20	M4	29	0.808
20	M4	30	–0.056
21	F1	3	0.76
21	F1	4	0.472
21	F1	5	0.76
21	F1	12	0.712
21	F1	16	0.04
21	F1	17	0.088
21	F1	24	0.136
21	F1	25	0.232
21	F1	30	0.184
21	F3	1	0.952
21	F3	8	–0.008

Table B-1. (Continued).

Listener	Talker	Condition	RA
21	F3	9	0.04
21	F3	10	0.712
21	F3	18	- 0.008
21	F3	23	0.328
21	F3	26	0.088
21	F3	27	- 0.008
21	F3	28	0.76
21	F3	29	0.808
21	F4	2	0.424
21	F4	6	0.04
21	F4	7	0.184
21	F4	11	0.52
21	F4	13	0.136
21	F4	14	0.184
21	F4	15	0.136
21	F4	19	0.712
21	F4	20	0.472
21	F4	21	0.904
21	F4	22	0.184
21	M1	3	0.856
21	M1	4	0.232
21	M1	5	0.376
21	M1	12	0.52
21	M1	16	– 0.008
21	M1	17	– 0.008
21	M1	24	0.04
21	M1	25	– 0.056
21	M1	30	0.28
21	M3	1	1
21	M3	8	– 0.056
21	M3	9	– 0.008
21	M3	10	0.808
21	M3	18	– 0.008
21	M3	23	0.088
21	M3	26	– 0.008
21	M3	27	0.088
21	M3	28	0.856
21	M3	29	0.856
21	M4	2	0.376
21	M4	6	0.088
21	M4	7	0.28
21	M4	11	0.28

Listener	Talker	Condition	RA
21	M4	13	– 0.008
21	M4	14	–0.008
21	M4	15	–0.104
21	M4	19	0.616
21	M4	20	0.52
21	M4	21	0.664
21	M4	22	0.28
22	F1	3	1
22	F1	4	0.328
22	F1	5	0.664
22	F1	12	0.808
22	F1	16	0.04
22	F1	17	0.04
22	F1	24	–0.008
22	F1	25	– 0.008
22	F1	30	0.232
22	F3	1	0.856
22	F3	8	–0.008
22	F3	9	0.184
22	F3	10	0.808
22	F3	18	–0.008
22	F3	23	0.424
22	F3	26	0.04
22	F3	27	0.04
22	F3	28	0.952
22	F3	29	0.808
22	F4	2	0.424
22	F4	6	0.136
22	F4	7	–0.008
22	F4	11	0.328
22	F4	13	0.136
22	F4	14	0.328
22	F4	15	– 0.008
22	F4	19	0.76
22	F4	20	0.472
22	F4	21	0.856
22	F4	22	0.088
22	M1	3	0.76
22	M1	4	0.28
22	M1	5	0.568
22	M1	12	0.472
22	M1	16	–0.008

Table B-1. (Continued).

Listener	Talker	Condition	RA
22	M1	17	–0.008
22	M1	24	–0.056
22	M1	25	–0.008
22	M1	30	0.04
22	M3	1	0.904
22	M3	8	–0.104
22	M3	9	–0.008
22	M3	10	0.424
22	M3	18	0.136
22	M3	23	0.28
22	M3	26	0.04
22	M3	27	–0.056
22	M3	28	0.952
22	M3	29	0.952
22	M4	2	0.232
22	M4	6	0.04
22	M4	7	0.088
22	M4	11	0.328
22	M4	13	0.088
22	M4	14	0.184
22	M4	15	–0.008
22	M4	19	0.616
22	M4	20	0.328
22	M4	21	0.568
22	M4	22	0.136
23	F1	6	0.136
23	F1	7	0.184
23	F1	9	0.184
23	F1	10	0.664
23	F1	11	0.376
23	F1	13	–0.056
23	F1	18	–0.008
23	F1	19	0.712
23	F1	21	0.808
23	F1	23	0.52
23	F1	27	0.088
23	F1	29	0.856
23	F3	2	0.376
23	F3	3	0.856
23	F3	5	0.472
23	F3	14	0.184
23	F3	15	–0.008

Listener	Talker	Condition	RA
23	F3	16	0.04
23	F3	20	0.424
23	F3	22	0.088
23	F3	24	–0.104
23	F3	30	0.088
23	F4	1	0.808
23	F4	4	0.088
23	F4	8	0.136
23	F4	12	0.52
23	F4	17	0.136
23	F4	25	–0.2
23	F4	26	–0.008
23	F4	28	0.76
23	M1	6	0.088
23	M1	7	0.232
23	M1	9	–0.008
23	M1	10	0.76
23	M1	11	0.136
23	M1	13	0.104
23	M1	18	0.088
23	M1	19	0.712
23	M1	21	0.376
23	M1	23	0.28
23	M1	27	–0.008
23	M1	29	0.664
23	M3	2	0.28
23	M3	3	0.664
23	M3	5	0.328
23	M3	14	0.184
23	M3	15	0.04
23	M3	16	0.04
23	M3	20	0.184
23	M3	22	–0.152
23	M3	24	–0.152
23	M3	30	–0.15
23	M4	1	0.904
23	M4	4	0.04
23	M4	8	0.088
23	M4	12	0.424
23	M4	17	–0.008
23	M4	25	–0.008
23	M4	26	0.04

Table B-1. (Continued).

Listener	Talker	Condition	RA
23	M4	28	0.856
24	F1	6	–0.104
24	F1	7	–0.056
24	F1	9	–0.008
24	F1	10	0.856
24	F1	11	0.28
24	F1	13	0.136
24	F1	18	–0.104
24	F1	19	0.952
24	F1	21	0.808
24	F1	23	0.184
24	F1	27	–0.008
24	F1	29	0.904
24	F3	2	0.76
24	F3	3	0.808
24	F3	5	0.376
24	F3	14	–0.008
24	F3	15	–0.008
24	F3	16	–0.104
24	F3	20	0.616
24	F3	22	0.088
24	F3	24	0.04
24	F3	30	0.136
24	F4	1	0.76
24	F4	4	–0.008
24	F4	8	0.232
24	F4	12	0.616
24	F4	17	0.04
24	F4	25	0.04
24	F4	26	0.136
24	F4	28	1
24	M1	6	–0.104
24	M1	7	–0.104
24	M1	9	0.04
24	M1	10	0.808
24	M1	11	0.328
24	M1	13	–0.008
24	M1	18	–0.008
24	M1	19	0.712
24	M1	21	0.472
24	M1	23	0.088
24	M1	27	0.088

Listener	Talker	Condition	RA
24	M1	29	0.712
24	M3	2	0.472
24	M3	3	0.856
24	M3	5	0.472
24	M3	14	–0.056
24	M3	15	0.04
24	M3	16	–0.104
24	M3	20	0.424
24	M3	22	0.136
24	M3	24	–0.056
24	M3	30	–0.056
24	M4	1	0.904
24	M4	4	0.04
24	M4	8	0.088
24	M4	12	0.376
24	M4	17	0.04
24	M4	25	–0.008
24	M4	26	–0.008
24	M4	28	0.76
25	F1	2	0.52
25	F1	3	0.76
25	F1	7	0.424
25	F1	8	0.232
25	F1	9	0.04
25	F1	10	0.808
25	F1	11	0.616
25	F1	14	0.424
25	F1	16	0.232
25	F1	17	0.136
25	F1	20	0.52
25	F1	24	–0.056
25	F1	30	0.04
25	F3	1	0.856
25	F3	5	0.568
25	F3	6	–0.008
25	F3	12	0.472
25	F3	13	0.04
25	F3	18	0.088
25	F3	19	0.856
25	F3	21	0.664
25	F3	22	0.088
25	F3	23	0.136

Table B-1. (Continued).

Listener	Talker	Condition	RA
25	F3	29	0.808
25	F4	4	0.088
25	F4	15	–0.008
25	F4	25	–0.2
25	F4	26	0.136
25	F4	27	0.136
25	F4	28	0.808
25	M1	2	0.328
25	M1	3	0.76
25	M1	7	–0.104
25	M1	8	0.28
25	M1	9	0.136
25	M1	10	0.856
25	M1	11	0.136
25	M1	14	0.136
25	M1	16	–0.152
25	M1	17	–0.056
25	M1	20	0.28
25	M1	24	0.328
25	M1	30	–0.056
25	M3	1	0.856
25	M3	5	0.616
25	M3	6	–0.056
25	M3	12	0.568
25	M3	13	–0.008
25	M3	18	–0.008
25	M3	19	0.664
25	M3	21	0.424
25	M3	22	0.088
25	M3	23	–0.056
25	M3	29	0.76
25	M4	4	0.232
25	M4	15	0.088
25	M4	25	–0.104
25	M4	26	0.04
25	M4	27	–0.056
25	M4	28	0.952
26	F1	2	0.76
26	F1	3	0.664
26	F1	7	0.376
26	F1	8	0.424
26	F1	9	–0.008

Listener	Talker	Condition	RA
26	F1	10	0.904
26	F1	11	0.424
26	F1	14	0.28
26	F1	16	0.136
26	F1	17	0.184
26	F1	20	0.808
26	F1	24	0.136
26	F1	30	0.184
26	F3	1	0.76
26	F3	5	0.664
26	F3	6	0.328
26	F3	12	0.76
26	F3	13	–0.056
26	F3	18	0.136
26	F3	19	0.76
26	F3	21	0.616
26	F3	22	0.184
26	F3	23	0.184
26	F3	29	0.808
26	F4	4	0.184
26	F4	15	–0.152
26	F4	25	–0.056
26	F4	26	0.28
26	F4	27	0.28
26	F4	28	0.952
26	M1	2	0.424
26	M1	3	0.856
26	M1	7	–0.008
26	M1	8	0.04
26	M1	9	0.232
26	M1	10	0.76
26	M1	11	0.472
26	M1	14	0.088
26	M1	16	–0.104
26	M1	17	–0.056
26	M1	20	0.568
26	M1	24	0.04
26	M1	30	0.136
26	M3	1	0.856
26	M3	5	0.568
26	M3	6	–0.008
26	M3	12	0.232

Table B-1. (Continued).

Listener	Talker	Condition	RA
26	M3	13	0.088
26	M3	18	0.04
26	M3	19	0.616
26	M3	21	0.52
26	M3	22	–0.056
26	M3	23	0.136
26	M3	29	0.616
26	M4	4	0.184
26	M4	15	0.04
26	M4	25	–0.00
26	M4	26	0.328
26	M4	27	–0.056
26	M4	28	0.952
27	F1	6	0.136
27	F1	15	–0.056
27	F1	18	–0.104
27	F1	22	0.088
27	F1	25	–0.056
27	F1	26	0.04
27	F3	2	0.472
27	F3	4	0.04
27	F3	11	0.52
27	F3	14	0.136
27	F3	16	–0.008
27	F3	27	0.04
27	F3	28	0.952
27	F3	30	0.04
27	F4	1	0.856
27	F4	3	0.856
27	F4	5	0.664
27	F4	7	0.04
27	F4	8	0.184
27	F4	9	–0.104
27	F4	10	0.952
27	F4	12	0.616
27	F4	13	0.136
27	F4	17	0.088
27	F4	19	0.664
27	F4	20	0.472
27	F4	21	0.568
27	F4	23	0.04
27	F4	24	–0.056

Listener	Talker	Condition	RA
27	F4	29	0.664
27	M1	6	–0.104
27	M1	15	0.04
27	M1	18	0.04
27	M1	22	0.088
27	M1	25	–0.104
27	M1	26	–0.008
27	M3	2	0.328
27	M3	4	0.136
27	M3	11	0.088
27	M3	14	0.232
27	M3	16	0.04
27	M3	27	0.04
27	M3	28	0.856
27	M3	30	–0.056
27	M4	1	0.904
27	M4	3	0.712
27	M4	5	0.568
27	M4	7	0.184
27	M4	8	–0.056
27	M4	9	–0.008
27	M4	10	0.904
27	M4	12	0.424
27	M4	13	–0.056
27	M4	17	–0.056
27	M4	19	0.904
27	M4	20	0.568
27	M4	21	0.664
27	M4	23	0.184
27	M4	24	–0.008
27	M4	29	0.952
28	F1	6	0.088
28	F1	15	–0.008
28	F1	18	–0.152
28	F1	22	0.184
28	F1	25	0.04
28	F1	26	0.04
28	F3	2	0.664
28	F3	4	0.088
28	F3	11	0.28
28	F3	14	0.136
28	F3	16	0.28

Table B-1. (Continued).

Listener	Talker	Condition	RA
28	F3	27	0.088
28	F3	28	0.904
28	F3	30	–0.008
28	F4	1	0.808
28	F4	3	0.904
28	F4	5	0.52
28	F4	7	0.04
28	F4	8	0.04
28	F4	9	0.04
28	F4	10	0.808
28	F4	12	0.568
28	F4	13	0.04
28	F4	17	0.184
28	F4	19	0.76
28	F4	20	0.52
28	F4	21	0.472
28	F4	23	0.52
28	F4	24	–0.152
28	F4	29	0.712
28	M1	6	–0.056
28	M1	15	–0.008
28	M1	18	–0.008
28	M1	22	0.088
28	M1	25	–0.056
28	M1	26	–0.056
28	M3	2	0.424
28	M3	4	0.088
28	M3	11	0.136
28	M3	14	–0.008
28	M3	16	–0.104
28	M3	27	0.04
28	M3	28	0.856
28	M3	30	–0.056
28	M4	1	0.856
28	M4	3	0.904
28	M4	5	0.664
28	M4	7	0.04
28	M4	8	0.328
28	M4	9	–0.008
28	M4	10	0.712
28	M4	12	0.424
28	M4	13	–0.152

Listener	Talker	Condition	RA
28	M4	17	0.04
28	M4	19	0.808
28	M4	20	0.472
28	M4	21	0.616
28	M4	23	–0.008
28	M4	24	–0.008
28	M4	29	0.712
29	F1	1	0.952
29	F1	4	0.472
29	F1	5	0.52
29	F1	12	0.616
29	F1	13	0.232
29	F1	19	1
29	F1	21	0.568
29	F1	23	0.52
29	F1	27	0.424
29	F1	28	0.904
29	F1	29	0.904
29	F3	3	0.904
29	F3	7	0.136
29	F3	8	0.136
29	F3	9	0.088
29	F3	10	0.808
29	F3	15	0.184
29	F3	17	0.184
29	F3	20	0.376
29	F3	24	0.136
29	F3	25	0.04
29	F3	26	0.184
29	F4	2	0.136
29	F4	6	0.136
29	F4	11	0.424
29	F4	14	0.04
29	F4	16	–0.008
29	F4	18	–0.056
29	F4	22	0.184
29	F4	30	0.04
29	M1	1	0.904
29	M1	4	0.184
29	M1	5	0.808
29	M1	12	0.328
29	M1	13	–0.008

Table B-1. (Continued).

Listener	Talker	Condition	RA
29	M1	19	0.712
29	M1	21	0.52
29	M1	23	0.136
29	M1	27	–0.008
29	M1	28	0.952
29	M1	29	0.76
29	M3	3	0.616
29	M3	7	0.088
29	M3	8	0.088
29	M3	9	0.088
29	M3	10	0.904
29	M3	15	–0.008
29	M3	17	0.088
29	M3	20	0.088
29	M3	24	–0.2
29	M3	25	0.088
29	M3	26	–0.152
29	M4	2	0.664
29	M4	6	0.088
29	M4	11	0.232
29	M4	14	0.136
29	M4	16	–0.008
29	M4	18	0.04
29	M4	22	0.088
29	M4	30	0.136
30	F1	1	0.904
30	F1	4	0.136
30	F1	5	0.472
30	F1	12	0.616
30	F1	13	0.28
30	F1	19	0.904
30	F1	21	0.616
30	F1	23	0.28
30	F1	27	0.184
30	F1	28	0.856
30	F1	29	0.664
30	F3	3	0.904
30	F3	7	–0.008
30	F3	8	0.28
30	F3	9	–0.056
30	F3	10	0.856
30	F3	15	0.088

Listener	Talker	Condition	RA
30	F3	17	0.28
30	F3	20	0.424
30	F3	24	0.136
30	F3	25	0.088
30	F3	26	0.184
30	F4	2	0.376
30	F4	6	0.088
30	F4	11	0.472
30	F4	14	0.088
30	F4	16	0.088
30	F4	18	–0.008
30	F4	22	0.088
30	F4	30	0.088
30	M1	1	0.952
30	M1	4	0.136
30	M1	5	0.568
30	M1	12	0.376
30	M1	13	–0.008
30	M1	19	0.904
30	M1	21	0.472
30	M1	23	0.232
30	M1	27	–0.104
30	M1	28	0.904
30	M1	29	0.76
30	M3	3	0.52
30	M3	7	0.088
30	M3	8	0.04
30	M3	9	–0.056
30	M3	10	0.664
30	M3	15	0.04
30	M3	17	–0.056
30	M3	20	0.136
30	M3	24	–0.104
30	M3	25	–0.056
30	M3	26	–0.008
30	M4	2	0.52
30	M4	6	–0.056
30	M4	11	0.28
30	M4	14	–0.008
30	M4	16	0.04
30	M4	18	–0.104
30	M4	22	0.232
30	M4	30	0.04

APPENDIX C: SOURCE SPEECH RECORDING INFORMATION

Some speech material in this test was recorded in the laboratory at the Institute for Telecommunication Sciences in Boulder, Colorado. Recordings were performed in an NC-35-rated sound isolation chamber according to the following.

The microphone used in the source recording was a Shure Beta 53A microphone sampled at 48 kHz/16 bit on a Windows-based computer using commercially available software. Active signal level was normalized to -28 dB below overload using the ITU-T Recommendation P.56 voltmeter software [6] [11].

The speech material spoken by the talkers was the word list defined in the MRT description of [3] in the carrier sentence, "Please select the word ...."

APPENDIX D: VOICE CODER SHORTHAND NOTATION

Table D- 1. Voice Coder Shorthand Notation Used by TIA.

New Name	Modulation	Name in 2003	Gross b/s	Net b/s	FEC	Back Ground Noise	Tones
QFA	QPSK-c	Baseline	7200	4400	Hard Dec		
QHA	QPSK-c	HR	3600	2450	Soft Dec		
FFB	F4FM	--	6300	4400	Soft Dec	X	X
FHB	F4FM	--	6300	2450	Soft Dec	X	X
QDB	QPSK-c	--	7200 / 3600	4400 / 2450	Soft Dec	X	X
QFB	QPSK-c	EFR	7200	4400	Soft Dec	X	X
QHB	QPSK-c	EHR	3600	2450	Soft Dec	X	X
QHC	QPSK-c	EHRS	3300	2250	Soft Dec	X	X

First Character	Modulation
Q	QPSK-c modulation, includes C4FM and CQPSK
F	F4FM modulation
Second Character	Full Rate/Half Rate/Dual Rate
F	Full Rate
H	Half Rate
D	Dual Rate (Full and Half)
Third Character	Suffix Enhancement/Bit Stealing
A	Basic, not enhanced
B	Enhanced
C	Enhanced with bit stealing

APPENDIX E: LISTENING LABORATORY CONFIGURATION

Two test chambers were set up to meet the standards set forth in Sections 8.10.4.10 - 8.10.4.15 of [9]. The physical layout of the chambers can be seen in Figure E-1. The loudspeaker carrying the speech signal sat on a table and was placed equidistant from the chamber side walls, on the edge of the table nearest the listener. The listening position was equidistant from the chamber side walls, and 150 cm away from the speech loudspeaker (in analogy to the talker-listener distance specified in 8.10.4.10). The two loudspeakers on either side of the table were used to produce the pink noise (allowed by 8.10.4.14). The loudspeakers were pointed toward the "back" of the room, and were not pointed directly at the listener, thus fulfilling 8.10.4.12. The combination of using two loudspeakers to produce the pink noise and the distance from the loudspeakers to the listening position created a quasi-uniform field of sound, thus satisfying 8. 10.4. 13.

In order to generate a field of "broadband pink noise" in compliance with section 8.10.4.11 of [9], a Gold Line Model PN2 Pink Noise Generator (PNG) was used. The PNG was modified to accept an external power source so any possible effect of a non-constant battery voltage could be avoided. The output of the PNG was fed into a General Radio Model 1952 Universal Filter, which was tuned to have a bandpass characteristic for the interval between 400 Hz and 4 kHz. The signal was then split into two signals in order to supply two different test chambers. After splitting, each signal passed through a mixer, equalizer, power amplifier and was finally delivered to the pair of loudspeakers located in a chamber. A Gold Line Model DSP30RM realtime spectrum analyzer was used to analyze the resulting acoustic pink noise spectrum in each chamber. Using information from the spectrum analyzer, the equalizers were tuned to achieve the best possible response for each test chamber.

Section 8.10.4.11 specifies that the pink noise generated should have a tolerance of 6 dB per octave band in the range of frequencies between 400 Hz and 4 kHz. In this range there are eleven third-octave bands. This means that only three full octave bands can be measured in this range, yet three octave bands do not fully cover the region of interest. However, it is possible to analyze two different sets of three full octave bands by grouping third-octave bands as shown in Tables E-1 and E-2. These two sets of bands are called octave band set A (OBSA) and octave band set B (OBSB).

After equalization was performed for each experimental chamber, the pink noise in each chamber was recorded and analyzed. A third octave analysis was performed in MATLAB. Energy in OBSA and OBSB were calculated by calculating the sum of the three third-octave bands included in their respective octave band. The relative noise power results for OBSA and OBSB are shown in Table E-1 and Table E-2 respectively for each chamber. For each chamber the 0 dB reference point is the midpoint between the highest and lowest measured octave band noise power. These tables show that 8.10.4.11 is satisfied in all cases.

A Brüel and Kjær Model 2250 sound level meter was used to verify that the noise level met specifications. The noise level in Chamber 1 was measured to be 69.4 dBA, and the noise level in Chamber 2 was measured to be 69.7 dBA, fulfilling 8.10.4.15.

Speech was generated in the two chambers using Fostex 6301B loudspeakers. The speech signal originates in MATLAB, propagates through the PC's sound card, then to a mixer, and finally to the loudspeaker. When the signal path was active but no signal was being sent to the speakers, the noise level in Chamber 1 fell to 22.3 dBA, and the level in Chamber 2 fell to

20.3 dBA. When speech was active in Chamber 1, speech levels ranged from 67-88 dBA, with undistorted, noise-free speech registering around 82 dBA. Similarly, active speech levels in Chamber 2 ranged from 70-86 dBA, and undistorted, noise-free speech registered around 82 dBA.

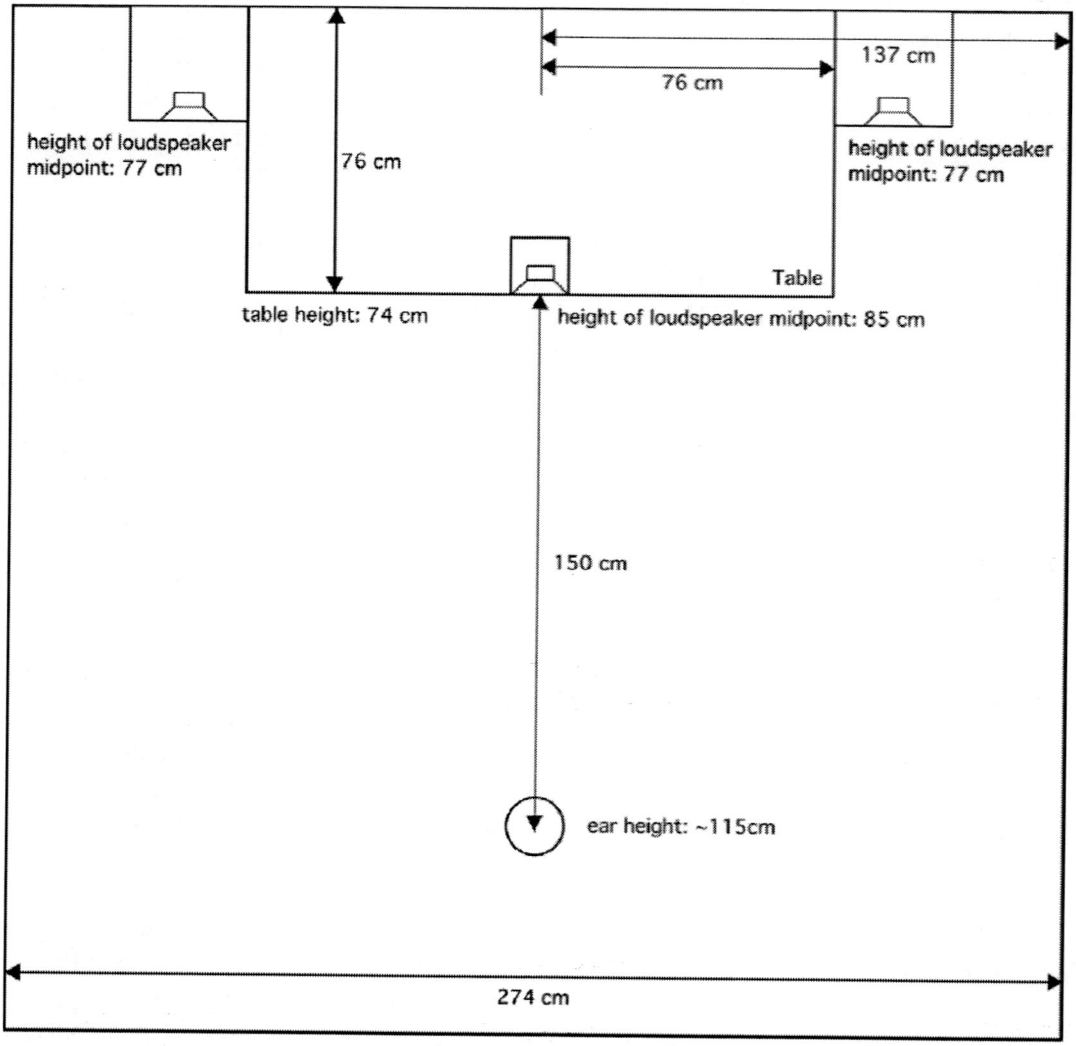

Figure E- 1. Listening laboratory layout.

Table E-1. Relative RMS Noise Power per Octave in Two Chambers Using OBSA.

Third Octave Bands	Chamber 1	Chamber 2
400, 500, 630 Hz	+4.1 dB	+4.0 dB
800,1000, 1250 Hz	+1.6 dB	+2.9 dB
1600, 2000, 2500 Hz	-0.5 dB	+1.3 dB

Table E-2. Relative RMS Noise Power per Octave in Two Chambers Using OBSB.

OBSB	Chamber 1 (in dB RMS)	Chamber 2 (in dB RMS)
630, 800, 1000 Hz	+2.1	+3.0
1250, 1600, 2000 Hz	+0.2	+2.4
2500, 3150, 4000 Hz	-4.2	-4

FORM **NTIA-29** (4-80)

U.S. DEPARTMENT OF COMMERCE
NAT'L. TELECOMMUNICATIONS AND INFORMATION ADMINISTRATION

BIBLIOGRAPHIC DATA SHEET

1. PUBLICATION NO. TR-08-453	2. Government Accession No.	3. Recipient's Accession No.
4. TITLE AND SUBTITLE Intelligibility of Selected Radio Systems in the Presence of Fireground Noise: Test Plan and Results		5. Publication Date June 2008 6. Performing Organization Code
7. AUTHOR(S) David J. Atkinson, Andrew A. Catellier		9. Project/Task/Work Unit No. 3141000-300
8. PERFORMING ORGANIZATION NAME AND ADDRESS Institute for Telecommunication Sciences National Telecommunications & Information Administration U.S. Department of Commerce 325 Broadway Boulder, CO 80305		10. Contract/Grant No.
11. Sponsoring Organization Name and Address National Telecommunications & Information Administration Herbert C. Hoover Building 14th & Constitution Ave., NW Washington, DC 20230		12. Type of Report and Period Covered
14. SUPPLEMENTARY NOTES		

15. ABSTRACT (A 200-word or less factual summary of most significant information. If document includes a significant bibliography or literature survey, mention it here.)

This report describes an experiment conducted to measure the intelligibility of selected radio communication systems when those systems are employed in high-background-noise environments experienced by firefighters. The test plan for a Modified Rhyme Test (MRT) is detailed, including requirements for source material preparation and listening test conduct. Finally, the results of the test, along with the data analysis, are presented. The results indicate that in some environments analog radios performed better than digital radios, and in some environments no radios performed well. This information should be considered whenever an agency is preparing to purchase and deploy a new communications system.

16. Key Words (Alphabetical order, separated by semicolons)

Intelligibility; Project 25; vocoder; modified rhyme test; noise; analog FM; land mobile radio; LMR, public safety, fire service

17. AVAILABILITY STATEMENT ☐ UNLIMITED.	18. Security Class. (This report) Unclassified	20. Number of pages 87
	19. Security Class. (This page) Unclassified	21. Price:

NTIA FORMAL PUBLICATION SERIES

NTIA MONOGRAPH (MG)

A scholarly, professionally oriented publication dealing with state-of-the-art research or an authoritative treatment of a broad area. Expected to have long-lasting value.

NTIA SPECIAL PUBLICATION (SP)

Conference proceedings, bibliographies, selected speeches, course and instructional materials, directories, and major studies mandated by Congress.

NTIA REPORT (TR)

Important contributions to existing knowledge of less breadth than a monograph, such as results of completed projects and major activities. Subsets of this series include:

NTIA RESTRICTED REPORT (RR)

Contributions that are limited in distribution because of national security classification or Departmental constraints.

NTIA CONTRACTOR REPORT (CR)

Information generated under an NTIA contract or grant, written by the contractor, and considered an important contribution to existing knowledge.

JOINT NTIA/OTHER-AGENCY REPORT (JR)

This report receives both local NTIA and other agency review. Both agencies' logos and report series numbering appear on the cover.

NTIA SOFTWARE & DATA PRODUCTS (SD)

Software such as programs, test data, and sound/video files. This series can be used to transfer technology to U.S. industry.

NTIA HANDBOOK (HB)

Information pertaining to technical procedures, reference and data guides, and formal user's manuals that are expected to be pertinent for a long time.

NTIA TECHNICAL MEMORANDUM (TM)

Technical information typically of less breadth than an NTIA Report. The series includes data, preliminary project results, and information for a specific, limited audience.

End Notes

[1] The authors are with the Institute for Telecommunication Sciences, National Telecommunications and Information Administration, U.S. Department of Commerce, Boulder, CO 80305.

[2] Appendix D contains a description of the notation used to identify codecs being tested in the TIA/Project 25 community. In this case, the notation means: QPSK-c modulation ("Q"), full rate ("F"), baseline ("A").

[3] Plus attenuation of the signal due to the mask. This is approximately 9 dB.

[4] [9] specifies an unbalanced talker pool of 4 male and 1 female talkers. To provide more gender-balanced results that could have wider applicability, a balanced talker pool of 3 male and 3 female talkers is used.

[5] dBC refers to a specific weighting function applied to a sound pressure level measurement. The "C" weighting function includes a broader spectrum of sound waves in the computation than the "A" weighting function. The "A" weighting function emphasizes the speech band and measurements using this function are measured in dBA.

[6] Note that only 45,000 of the 54,000 sentences will be used. This corresponds to 5 words from each block of 6 words, defined in [3], for each talker for each condition. Selections from the extra words are used for training.

[7] While the actual value of the 25 kHz Analog FM system was 78%, this was statistically equivalent to 80% as computed by the T-test defined in [9].

In: Intelligibility Research and Communication Issues… ISBN: 978-1-61668-634-5
Editor: Samuel A. Fletcher

Chapter 2

Relationships between Intelligibility, Speaker Identification, and the Detection of Dramatized Urgency*

***A. Andrew Catellier and D. Stephen Voran*[1]**

Abbreviations/Acronyms

DOC	Department of Commerce
DU	Dramatized Urgency
IP	Internet Protocol
ITS	Institute for Telecommunication Sciences
LAN	Local Area Network
LDC	Linguistic Data Consortium
MNRU	Modulated Noise Reference Unit
NTIA	National Telecommunications and Information Administration
NTP	Normalized Task Performance
PCM	Pulse Code Modulation
PDA	Personal Digital Assistant
SID	Speaker Identification
SNR	Signal-to-Noise Ratio
SUSAS	Speech Under Simulated and Actual Stress
TIS	Task Induced Stress
TSID	Tactical Speaker Identification Database

The systems used for public safety speech communications must be intelligible. It is also desirable that they transmit secondary information, such as the attributes of a speaker's voice.

* This is an edited, reformatted and augmented version of a U. S. Department of Commerce publication dated November 2008.

[1] The authors are with the Institute for Telecommunication Sciences, National Telecommunications and Information Administration, U.S. Department of Commerce, Boulder, CO 80305.

This secondary information can allow a user to identify the speaker and his or her emotional state. Testing speech communications systems for the delivery of intelligible speech is common. Testing for human perception of the delivery of this secondary information is less common, though some prior work has been done. Building on this prior work, we describe a set of controlled laboratory listening experiments. These experiments characterize the relationships between speech intelligibility, speaker identification, and the detection of dramatized urgency in a speaker's voice across a range of simulated speech processing conditions. The experiment results indicate that for the speech processing conditions considered here, detection of dramatized urgency is the most robust property, speaker identification is less robust, and speech intelligibility is the least robust.

Keywords: human listening tests; intelligible speech; speaker identification; speaker stress detection; speaker urgency detection; speech transmission system; subjective speech quality tests

1. Introduction

Public safety speech communication systems are designed to carry a message from a speaking user (speaker) to a listening user (listener). The most important characteristic of these systems is intelligibility (i.e., whether the listener can understand the message). In addition to intelligibility, public safety communication systems should aim to successfully transmit secondary information, such as attributes of the speaker's voice. Secondary information allows the listener to identify the speaker or to recognize the emotional state of the speaker.

The ability to identify or confirm the identity of a speaker is called "speaker identification" (SID). SID can be particularly important to public safety officials who rapidly communicate with each other to accomplish time-critical emergency operations. If speakers can be identified implicitly based on transmitted attributes of their voices, the additional overhead associated with explicit identification can be avoided (e.g., "This is Officer Roberts speaking."). Even more valuable is the possibility of detecting whether a speaker has lied about their identity.

Similarly, public safety officials often need to monitor numerous transmissions with only partial attention while simultaneously performing other important tasks. If one of many different speakers associated with one of many different transmissions displays a shift in emotional state via his or her voice, detecting that shift can be very important. When such a shift is detected (e.g., from a neutral tone to a tone of urgency), the listening public safety officials will commit full attention to that specific speaking official to provide support and aid.

This report describes the design, implementation, analysis, and results of a set of three controlled laboratory listening experiments. Each experiment uses a set of six different processing conditions (e.g., speech coders) that spanned a broad range of perceptual quality. The first experiment characterized the relative intelligibility in a sentence context of each processing condition. In this document, if intelligibility is mentioned, it can be assumed that we mean "intelligibility in a sentence context." In the intelligibility experiment, the subjects listened to a sentence and attempted to repeat it verbatim. The second experiment characterized human SID performance for each processing condition. In this SID experiment, listeners learned to recognize six different speakers, and then attempted to recognize each speaker's processed voice. The third experiment characterized a listener's ability to detect dramatized urgency in processed speech. In this experiment, listeners attempted to detect one of two dramatized emotional states based on the voice characteristics of a speaker. These emotional states are "dramatized neutral" and "dramatized urgency" (DU), so we refer to this experiment as "detection of DU." Together, these three experiments characterized the relationships between speech intelligibility, SID, and the detection of DU across six speech processing conditions.

In the sections that follow, we describe related work previously conducted by other researchers. We then describe the various speech recordings used in the three experiments, the six speech processing conditions, the three experiment designs, the software used, and the main results obtained. The results show how intelligibility, SID, and detection of DU vary as a function of the six speech processing conditions used.

2. Previous Research

Much work has been done to develop means for testing speech communications systems. However, testing for human perception of the delivery of the secondary information is less common. We are not aware of any previous efforts to characterize how the human detection of speaker emotional state is influenced by the distortions caused by speech processing associated with communication systems. Prior work has been done on the related topics of automatic recognition of speaker emotions [1] and automatic speech recognition that is invariant to speaker emotions [2].

Various studies related to SID have been conducted over the years. In 1963 Compton studied human SID abilities for multiple filtered versions of the sustained vowel sound at the end of the word "three" [3]. He found that SID can happen with recordings as short as 1/40 of a second. He also found that when the pitch of different speakers was closer, those speakers were more easily confused.

Bricker and Pruzansky conducted an experiment where coworkers were asked to identify speakers using processed speech recordings. The speakers were familiar to the listeners, and pictures were used to aid the identification process [4].

Uzdy used two different low-rate vocoders to conduct a SID experiment where listeners were familiar with the speakers [5]. His goal was to determine each vocoder's effectiveness in transmitting data pertinent to SID. This goal is similar to our current work. Uzdy discussed the importance of adequate listener training and noted that about five hours of training were necessary to obtain stable results.

Schmidt-Nielsen did significant, sustained work on human and machine SID performance, SID performance for familiar and new speakers, and the relations between SID performance and speech coding distortions [6-10]. In [6] she suggests using a small number of speakers to keep within the restrictions of listener memory. Quatieri describes significant work relating machine SID to coding distortions in [11].

3. SPEECH RECORDINGS

This section describes the speech recordings used for each experiment. Example speech files from the DU portion of this experiment are available at <http://www.its.bldrdoc.gov/pub/audio/pubs talks.php>.

3.1. Intelligibility

There are numerous approaches to testing the intelligibility of speech. Some approaches require the identification of isolated unrelated words, or leading consonants of rhyming words. In public safety communications, as in much of naturally occurring spoken communications, groups of words provide context for each other and that can help a listener to properly understand some hard-to-hear words. We selected "word intelligibility in a sentence context" as a suitable way to assess intelligibility in the present application. To test word intelligibility in sentence context, we selected and recorded 20 sentences from current issues of *The Wall Street Journal* and *The New York Times*. Sentence lengths ranged from 6 to 14 words with a median length of 9 words (e.g., "This rebellion has forced banks to reduce bond offerings."). The sentences were selected to be of average complexity and to contain only commonly used words. The sentences are considered typical in terms of the amount of context the words within a sentence provide for each other. One female and one male speaker recorded each of the sentences. We used studio-grade digital recording equipment and a quiet recording room with average noise level below 20 dBA.

3.2. Speaker Identification

A search for North American English recordings to use in the SID experiment resulted in the selection of the Tactical Speaker Identification Database (TSID), which is available from the Linguistic Data Consortium (LDC) [14]. We chose this database because it includes semi-spontaneous speech, repeated utterances of lists of sentences and digits, and some utterances that are recorded by multiple speakers.

To ensure that the experiment size was manageable within the limits of human memory (as suggested in [8]), we decided to select three female speakers and three male speakers from the database. Average pitch and voicing strength were determined for each male and female speaker. We then looked for male speakers that spoke the same sentences and spanned the full range of pitches found in males in the database. Additional considerations in selecting male speakers and recordings included minimizing speaker script-reading errors, minimizing microphone handling and breath noises, and minimizing microphone overload distortion. We selected three of the four female speakers found in the database by maximizing the range of pitches and the quality of recordings.

After speaker selection, we looked for similar digit sequences (of lengths two and four) and sentences with similar content spoken by each speaker. These were used to form clips of three lengths: short, medium, and long. Semi-spontaneous speech was used for training purposes.

3.3. Detection of Speaker "Stress"

In general, we are interested in listener detection of speaker "stress." But the term "stress" is subjective, and covers a wide range of circumstances and resulting speech signals. For speech signals, objective refinement of the term "stress" and quantification of stressor levels is enabled through the use of known acoustic correlates. These include changes in level, tempo, pitch, and formants [1, 2, 13, 14].

3.3.1. Task induced stress

Previous work to develop and test automatic speech recognition that is invariant to speaker emotions resulted in the Speech under Simulated and Actual Stress (SUSAS) recorded speech database [13, 15]. One portion of the SUSAS database involved a male helicopter pilot recording isolated words in neutral (helicopter on the ground and running) and task (pilot flying helicopter) situations. Another portion included one male and one female speaker recording isolated words in neutral (no task) and computer-graphics based "dual tracking" task situations. When comparing the task recordings with the neutral recordings, only a minor sense of distraction is evident. We call this Task Induced Stress (TIS), and we used some of these recordings as the basis for a portion of this experiment. The experiment revealed that the detection of TIS is a very difficult task, independent of the speech processing condition. Because of this, the TIS portion of the experiment is not addressed in the remainder of this report.

3.3.2. Dramatized urgency

Since urgent speech is not uncommon in public safety communications, speakers conveying urgency were an important part of this experiment. It would not be ethical to subject speakers to events (e.g., physical dangers) that could create a true sense of urgency. Recording speakers confronted by naturally occurring urgency-inducing events was not a practical option, but this might be considered for potential future work. We elected to create recordings of DU.

We monitored public safety communication channels and transcribed messages between public safety personnel to use as scripts. Messages selected ranged in length from 2 words to 21 words with a median length of 9 words (e.g., "We have two children still trapped under the bus."). For comparison purposes, the scripts also included the isolated words of the TIS recordings.

One female and one male speaker recorded the DU scripts. We used studio-grade digital recording equipment and a quiet recording room with average noise level below 20 dBA. Each speaker read the scripts while verbally dramatizing two different situations: a non-urgent situation (neutral) and a situation requiring an urgent response (DU situation). We activated a set of rotating mirrored red and blue strobe lights to provide an unmistakable visual indication of when the speakers should dramatize urgency. A total of 16 different messages were used in the experiment.

3.3.3. Acoustic Correlates in Dramatized Urgency

We analyzed the DU recordings and identified several acoustic correlates. The level of DU speech increased (relative to neutral speech) by an average of 6.2 dB for the male speaker, and 8.0 dB for the female speaker. However, note that this level increase was not directly available to listeners because it was removed via level normalization. It may have been indirectly available if it was accompanied by audible sounds of increased speaking effort.

The two speakers responded oppositely in terms of tempo. The male speaker increased his talking tempo slightly in DU, so his average message duration decreased from 2.86 to 2.68 seconds. The female lengthened certain words for emphasis and thus decreased her tempo. Her average message duration increased from 2.73 to 3.01 seconds.

The mean pitch of the male speaker increased from 134 Hz (neutral) to 148 Hz (DU), while the standard deviation increased from 21 to 23 Hz. For the female speaker, the mean pitch increased from 211 to 249 Hz, and standard deviation increased from 18 to 38 Hz. The pitch histograms in Figure 1 show all four of these results. We also observed changes in formant structure for both speakers.

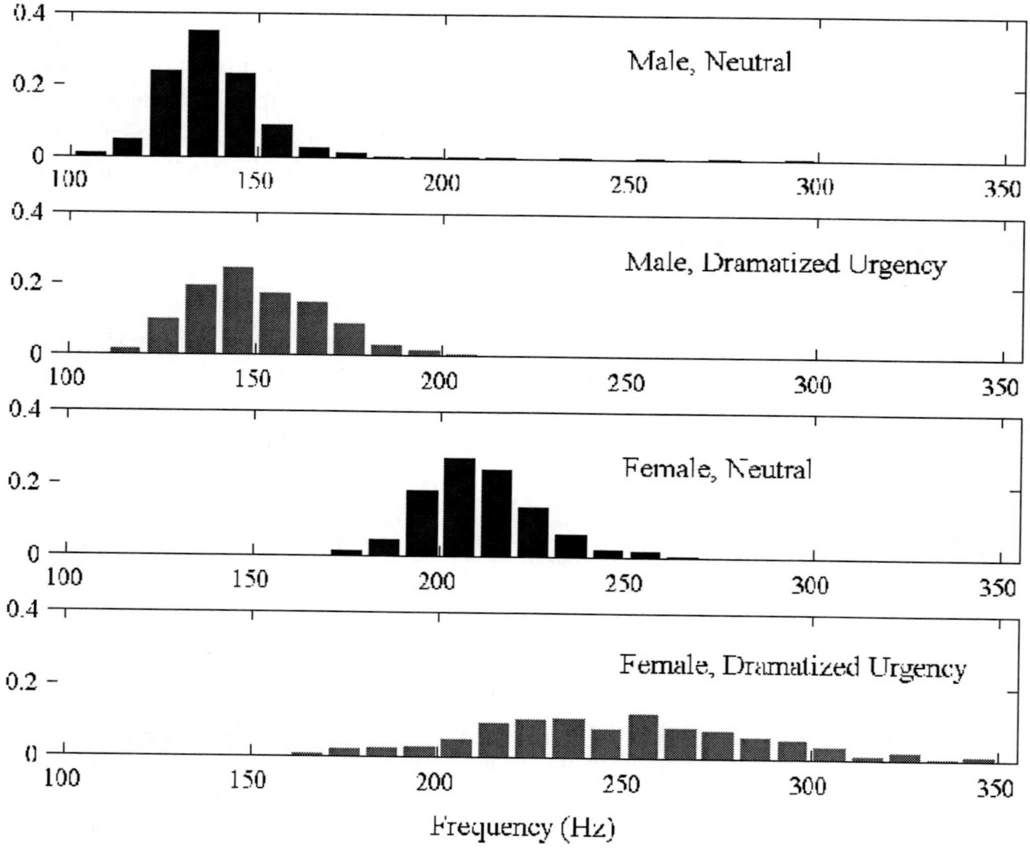

Figure 1. Pitch histograms for four cases as labeled.

The increases in mean pitch and pitch variation found in our DU recordings are qualitatively consistent with those found in cockpit voice recordings of a real stressful and urgent situation. These recordings document the voices of a pilot and copilot both when relaxed, and in the minutes before their aircraft crashes [14]. Whether or not DU is a good surrogate for true urgency will likely depend on numerous factors including individual speakers' physical and psychological characteristics and the details of the urgent situation.

All recordings for all experiments were resampled to a rate of 8,000 samples per second using the "PCM filter" option (160 to 3640-Hz bandpass filtering) provided in [16]. The level of each recording was then normalized to –26 dB, relative to clipping using tools from [16]. Next, the recordings were passed through software to implement various speech processing conditions.

4. Speech Processing Conditions

The goal of the experiments was to find the relationships between speech intelligibility, speaker identification, and the detection of dramatized urgency. The usefulness and robustness of these relationships is greatest when they span the widest possible range of these parameters. Thus, six speech processing conditions were chosen to provide the widest possible range of experimental results. Table 1 summarizes the six conditions.

C1 involves no additional processing and thus provides a best-case reference point for all three tasks. In C4, Modulated Noise Reference Unit (MNRU) [19] software produces multiplicative (speech-correlated) noise resulting in an active speech SNR of 6 dB. This is a standardized reference condition that can allow one to build relationships to other experiments that also include the MNRU.

The remaining conditions use three different narrowband (4-kHz nominal) speech codecs specified in standards or proposed standards for low bit-rate digital communication in the presence of acoustic background noise. These codecs simulate frequency-dependent voicing strength by adaptively mixing periodic and aperiodic excitation signals. Since communications are increasingly likely to travel through some IP network it is important to model and examine possible effects. For C6, three simulated communication systems are concatenated. The first and last are the same as C5 (speech encoding, bit errors in the transmission channel, then speech decoding). The middle system consists of packetization of the speech samples followed by the deletion of randomly selected packets and the insertion of an equal number of empty packets at different random locations. A packet loss concealment algorithm is used to extend previous speech samples into these inserted empty packets.

The speech processing conditions are certainly relevant to public safety speech communication systems. But evaluating the conditions is not the primary goal of these experiments. Rather the conditions are tools that enable the experiments to yield relationships between speech intelligibility, speaker identification, and the detection of dramatized urgency.

After creating recordings for each condition, the active speech level of each recording was again normalized to –26 dB relative to clipping.

Table 1. Six Conditions Used in the Speech Intelligibility, Speaker Identification, and Detection of Dramatized Urgency Experiments.

Condition (C)	Description
C1	Null (no further processing)
C2	Low rate speech coding
C3	Very low rate speech coding
C4	MNRU, Q = 6 dB SNR
C5	Low rate speech coding with bit errors
C6	C5+Severe Packetization Impairments+C5

5.1. Speech Intelligibility

Twenty-four randomly-selected listeners participated in the experiment. Sixteen were male, eight were female, estimated ages ranged from 20's to 60's with a mean estimated age of approximately 40, all were fluent in English, two reported slight hearing losses, and none were familiar with the technical details of the experiment. Listeners participated one at a time in a sound-isolated room where the average background noise level was below 20 dBA. The recordings were played over a powered monitor speaker with a single full-range four-inch driver. Listeners could adjust the listening level to their preferred level at any time.

In the experiment, listeners heard a recorded sentence and were asked to repeat it back. These responses were recorded and later evaluated for the number of correct words repeated. Listeners could not proceed until the entire sentence was played, and they were not allowed to replay any sentence. Progress through the experiment was controlled through a graphical interface on a PDA supported by a wireless LAN connection.

The experiment started with a practice session containing four trials. This session familiarized listeners with their task and with the procedures. Following this practice session, each listener then heard 24 sentences (4 per condition) and the sentences used with each condition were varied in a balanced way as the experiment progressed. The result was 96 intelligibility trials per condition (each sentence was used 4 times per condition, but only once per listener), for a total of 576 trials. Each listener heard the recordings in a different random order. After the experiment, several different statistical tests showed that no single listener was an outlier in this speech intelligibility task. Responses were recorded and later evaluated to compute the number of correct words repeated.

A second version of this experiment was later given to six additional listeners. This version used the same speech recordings, speech processing conditions, and general procedures. It differed from the original experiment only in that listeners were allowed to hear the recordings as many times as they wished. After each playing of a recording, listeners were asked to repeat the sentence as they heard it or to report that no words were understood. Responses were recorded and later evaluated for three quantities: the number of correct words repeated after the first playing, the number of correct words repeated after the final playing, and the total number of plays. On some occasions, a subject would fail to provide any response after a playing and this trial was scored as zero words correct.

This second version of the intelligibility experiment does not conform with typical approaches, but it does more closely parallel the SID and detection of DU experiments. In each of these, listeners were allowed to play recordings as many times as they wish.

5.2. Speaker Identification

The SID experiment design and procedures were refined several times using feedback from subjects who participated in early versions of the experiment. The final design included seven different parts. Three were actual experimental test sessions where data is collected, and four were supporting parts that were preliminary or tutorial in nature.

5.2.1. Listeners

Twenty-five randomly selected listeners participated in the experiment. Fifteen were male and ten were female. Their ages ranged from approximately 37 to 64 with a mean age of 49. None of the listeners were familiar with the technical details of the experiment. Listeners participated one at a time in a sound-isolated room where the average background noise level was below 20 dBA. Recordings were played over a powered monitor speaker with a single full-range four-inch driver. Listeners could adjust the listening level to their preferred level at any time throughout the experiment. The experiment, including all training and testing, took listeners from 45 to 90 minutes to complete, and the average completion time was just under one hour.

The randomly selected listener pool included two listeners with hearing aids and one listener who reported deafness in one ear. After careful consideration described in Section 6.2.1, we elected to include the data from these three listeners in the overall experiment results.

The experiment administrator received many hours of exposure to both undistorted and distorted recordings from the six speakers. After this incidental training, the experiment administrator also served as a listener. As described in Section 6.2.1, his results are not included in the overall experiment results.

5.2.2. Training and Quiz Session

The experiment starts with a session where the listener assigns a face and a name to each of the six different speakers. This session is provided to allow the experiment to better simulate the actual conditions under which listeners most often identify speakers that they cannot see. That is, listeners typically can reference a name, face, or both in memory when identifying speakers who cannot be seen.

To accomplish this goal, the SID experiment was controlled interactively through a computer program. Figure 2 shows the window presented during listener training. The middle of the window on the computer screen has one button for each speaker (e.g., "Speaker 1", "Speaker 2"). The listener presses a button associated with a speaker. The program plays several sentences spoken by this speaker, and allows the listener to select a name and portrait[2] (or identity) that seems appropriate for the speaker from the options listed on the left side of the screen. From now on, in addition to playing several sentences spoken by this speaker, the selected identity is displayed on the right side of the screen whenever the listener presses this speaker's button.

After hearing all six speakers and assigning a name and portrait to each, listeners take a short quiz. This quiz helps the listener check their knowledge of the speakers and become confident in their ability to recognize each speaker. The initial quiz state is shown in Figure 3. During the quiz, the listener hears a sentence. The listener identifies the speaker by clicking on the identity that he believes belongs to the currently playing speaker. The program immediately displays the selected identity prominently on the right side of the screen, as shown in Figure 4. If satisfied with the selection, the listener can confirm his choice; if not, he can make another selection. Upon confirmation, feedback indicating whether or not his selection is correct is displayed, as shown in Figure 5. The process repeats for all six speakers. Then the listener chooses to either continue training or proceed to the test sessions.

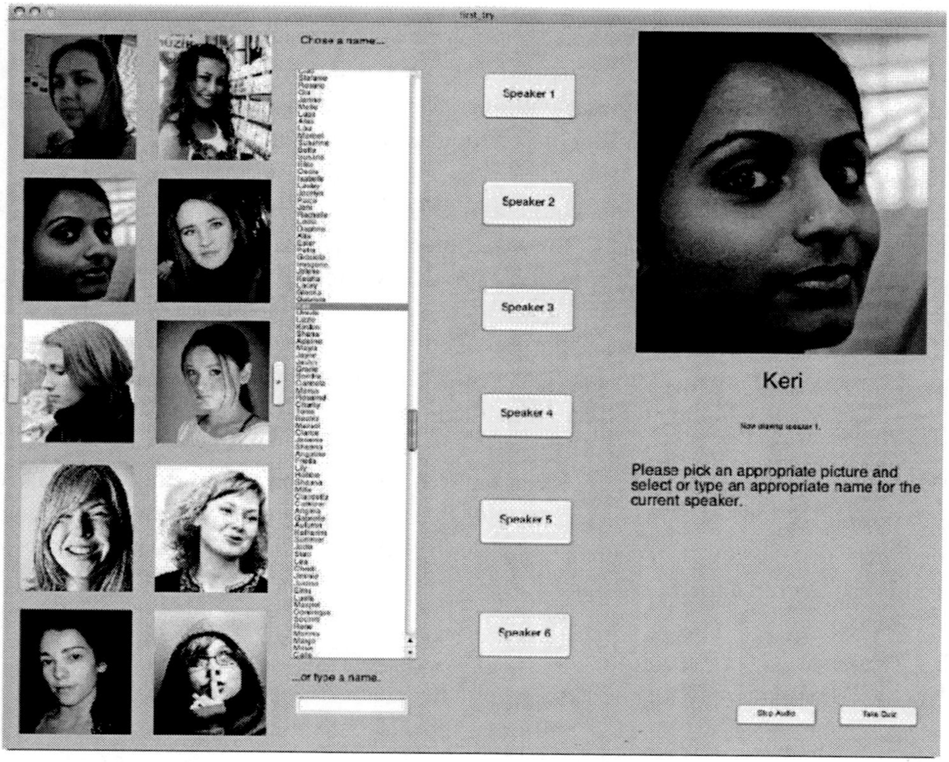

Figure 2. Speaker identification training interface for listeners.

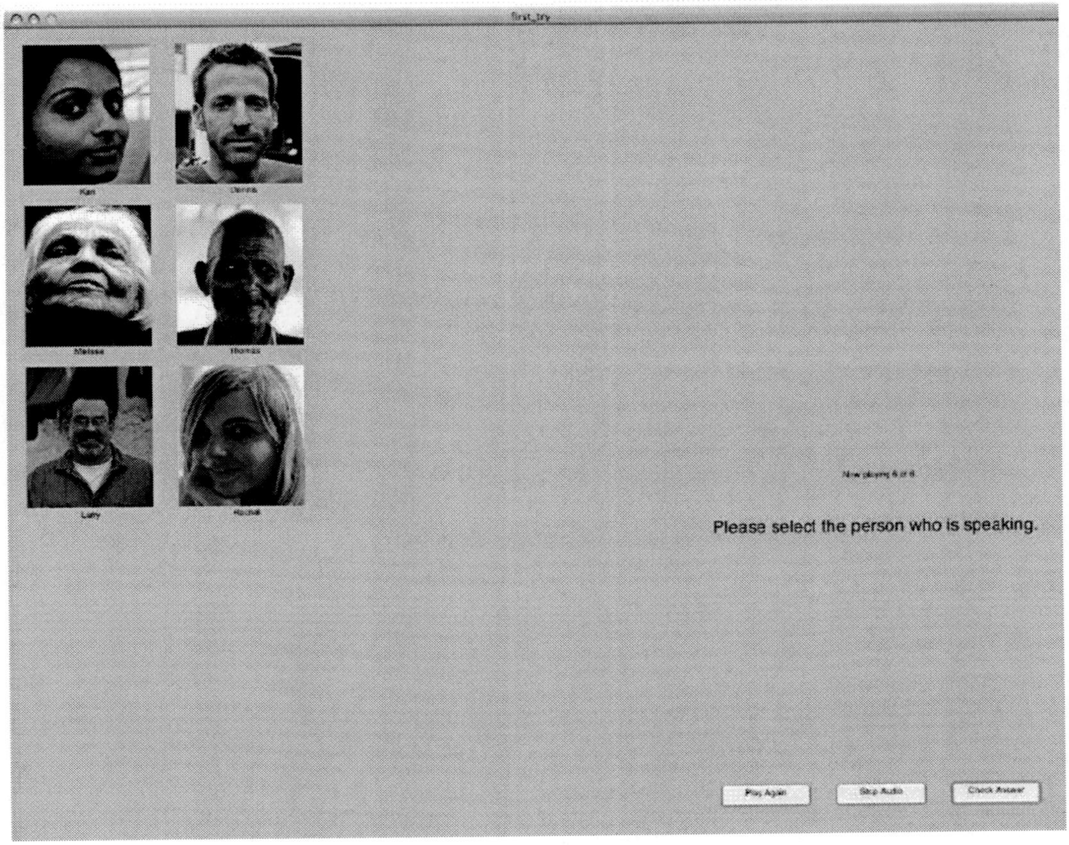

Figure 3. Speaker identification quiz interface for listeners—initial state.

5.2.3. Experimental Test Session 1—Sentences

The first experimental test session uses two sentences from each speaker. There are six speakers and six conditions for a total of 2 × 6 × 6 = 72 trials in the session. One sentence is the same for all six speakers: "Don't ask me to carry an oily rag like that." The second sentence differs for every speaker. The recordings used in this session range from approximately 1.7 to 2.6 seconds in length (8 to 13 syllables in length) with a mean value of about 2.2 seconds (about 11 syllables).

This experimental test session was presented to the listener in a fashion very similar to the aforementioned quiz session. One of the 72 available recordings was played back from the beginning of a randomized list. The randomized list was unique for each listener to prevent any potential order effects. The listener was asked to identify the speaker of the recording, and select the correct identity out of the six shown on the left side of the window, on a screen similar to Figure 3. Once clicked, the selected identity displayed prominently, similar to Figure 4, and the listeners were allowed to move on to the next recording. However, the listeners were allowed to select a different identity or replay the recording as many times as necessary. Unlike the quiz session, the listeners were not notified about the accuracy of their selection—the software simply moved on to the next recording in the randomized list after identity selection was confirmed.

Figure 4. Speaker identification quiz interface for listeners—after listener selection.

5.2.4. Experimental Test Sessions 2 and 3—Digits

A short reminder session is provided before experimental test sessions 2 and 3. The six chosen identities are displayed on the left side of the window. The listener is instructed to listen to each speaker at least once before moving on to the next experimental test session. By clicking any of the portraits, the listener can hear a recording of the corresponding speaker that is similar to those in the upcoming session. The listener can spend as much time in this reminder session as is desired, and the software requires that the listener listen to each speaker at least once. Once the reminder session is complete, experimental test sessions 2 and 3 are administered exactly as session 1 was.

The second experimental test session uses recordings that contain four spoken digits (e.g., "three six nine eight"). Each of the six speakers was associated with four recordings (of four digits each), so the session contained a total of $4 \times 6 \times 6 = 144$ trials. The recordings in this session range from approximately 1.3 to 1.9 seconds in length (4 to 5 syllables in length) with a mean value of about 1.6 seconds (about 4.4 syllables). There were 15 unique "4-digit" sequences—we did not control the recording of this database, so within the speakers only 15 unique combinations were available. With one exception, all speakers recorded a completely different sub-set of these 15 sequences. Pairs of these 4-digit sequences often have two or three digits in common. Because of these similarities and subtle differences, content-based SID would be extremely difficult (e.g., noticing which 4-digit sequences are associated with Speaker 1).

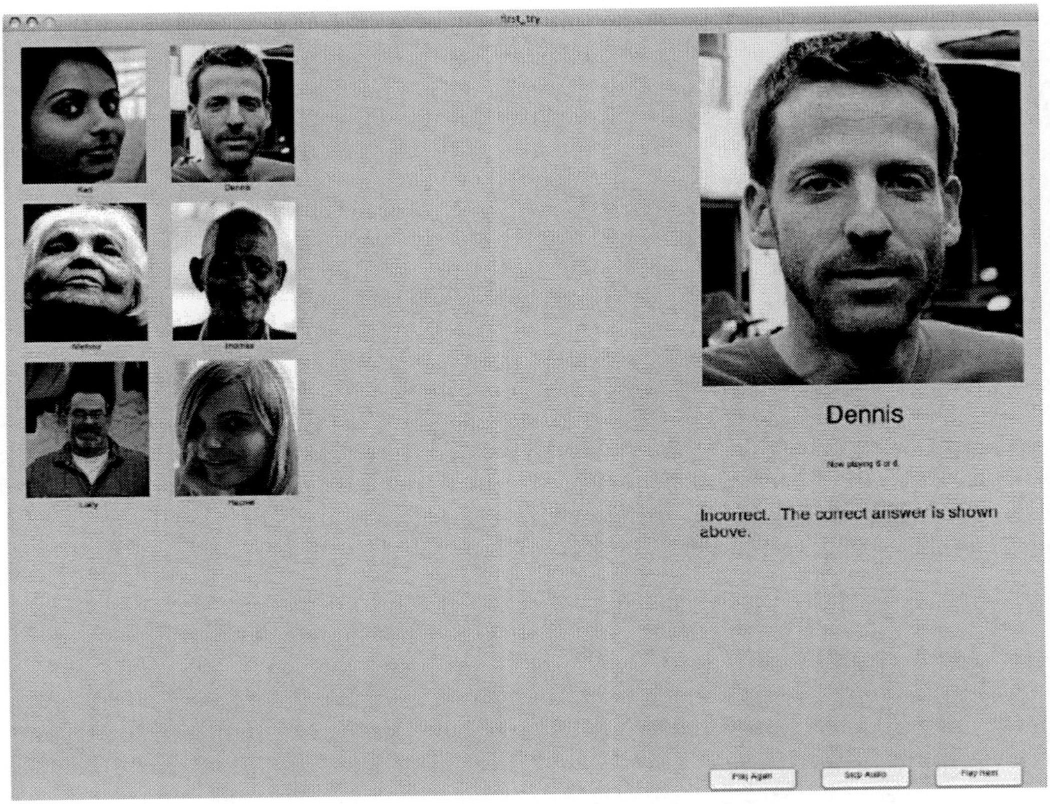

Figure 5. Speaker identification quiz interface for listeners—correct answer display.

The third session uses recordings that contain pairs of spoken digits. All six speakers provided the exact same four recordings: “five two”, “six zero”, “six three”, and “eight zero”. Thus, in this session, content is identical across speakers, and content-based SID is not possible. Here again, the session includes 144 trials. The recordings used in this section range from approximately 0.6 to 0.8 seconds in length (2 to 3 syllables in length) with a mean value of about 0.7 seconds (about 2.5 syllables).

The combined number of trials for all three experimental test sessions is 72 + 144 + 144 = 360.

5.3. Detection of Dramatized Urgency

This experiment used the same twenty-four randomly-selected listeners that participated in the speech intelligibility experiment (see Section 5.1). Recordings were played over a powered monitor speaker with a single full-range four-inch driver. Listeners could adjust the listening level to their preferred level at any time. Experiment progress and data collection were controlled through a graphical interface on a PDA supported by a wireless LAN connection. Listeners first participated in a practice session to familiarize them with the task and the procedure.

Listeners heard a recording and responded to the prompt "Please select the talker's stress or urgency level." Response options in each of these binary forced-choice trials were "Low" (the correct answer for neutral recordings) and "High" (the correct answer for TIS and DU recordings). Listeners could respond at any time once a recording had started to play, and could restart the playback at any time. In this manner, each listener could proceed at an individualized pace through the experiment. Each listener heard 192 trials for a total of 4,608 (192 trials × 24 listeners) DU detection data points.

6. RESULTS

Throughout this section and the next, we report results in terms of either fraction correctly identified or normalized task performance (NTP). We introduce the NTP scale because it enables a more direct comparison of the results from the three experiments. On the NTP scale, a value of 1 indicates perfect information from the listeners in the experiment, and a value of 0 indicates no information from the listeners. This is true for the intelligibility, SID, and detection of DU experiments.

6.1. Speech Intelligibility

In the intelligibility experiment, NTP is simply the fraction of words correctly repeated by a listener. If 100 percent of the words were repeated correctly, an NTP value of 1 would be the result. If none of the words were repeated correctly, then an NTP value of 0 would be the result.

Figure 6 shows the mean NTP values for each of the six speech processing conditions, after averaging over all listeners and all messages for each condition. The data points shown by the solid line describe the main intelligibility experiment (24 listeners, replaying recordings not allowed). The data points shown by the dashed and dotted lines correspond to results from the second version of the intelligibility experiment (six listeners, replaying recordings allowed). The dashed line shows results that were calculated after the first play and the dotted line shows results that were calculated after the final playing. The 95-percent confidence intervals (CI) for these three cases overlap and are omitted from this figure for clarity.

Figure 6 shows that allowing replays does improve NTP by a small amount (the dotted line is above the solid line). The larger increases in NTP are associated with cases of medium speech intelligibility (C4 and C5) where additional plays apparently can help at least some listeners with the task. The smaller increases in NTP are associated with cases of very high (C1, C2, and C3) and very low intelligibility (C6). It seems that in these limiting cases, additional plays provide limited advantage. The average number of plays generally increases with condition number from just over one (1.3 in C1) to two (2.0 in C6).

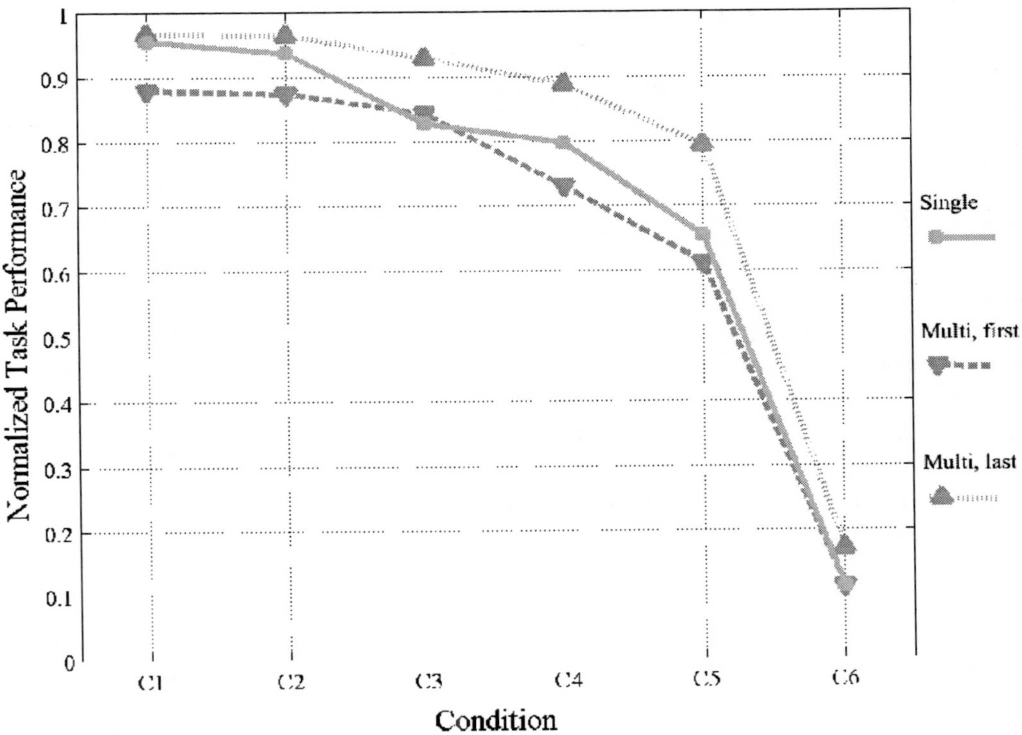

Figure 6. Intelligibility experiment: NTP values for three cases—solid line for single play experiment, dashed line for the first play, and dotted line for the last play in the multiplay experiment.

While allowing replays does improve NTP by a small amount this does not change the general trends as we move from C1 to C6. That is, all three lines in Figure 6 show a similar general trend: a gentle decrease in NTP followed by an abrupt drop. Based on this similarity of shape we argue that the final conclusions about relative robustness in Section 7 are not highly sensitive to the choice between these two intelligibility testing approaches (single play of each recording versus unlimited playing of each recording). The main intelligibility experiment (with no replays allowed) has much more data than the secondary intelligibility experiment (24 listeners vs. 6 listeners) and thus we focus on that experiment from this point forward in this report.

Figure 7 shows the mean NTP values and 95 percent CI for each of the six speech processing conditions, after averaging over all listeners and all messages for each condition. Note that as we move from C1 to C6, NTP drops monotonically from 0.95 to 0.11. This is an NTP drop of 0.84.

6.2. Speaker Identification

In the SID experiment, 360 trials were administered to 25 listeners in the main pool. This gives 9,000 data points. Each data point is one SID, which can be either correct or incorrect. Using this view, the data is binary in nature and can be modeled using the binomial distribution. In the binomial model, the maximum likelihood estimate for the probability of correct identification is simply as shown in (1).

$$\hat{p} = \frac{\textit{number of correct identifications}}{\textit{total number of identifications}} \tag{1}$$

The 95% CI for the estimate $\hat{p}$ is calculated as given in [18]. We report $\hat{p}$ as the fraction correctly identified in Section 6.2.1 and Section 6.2.2, and we report the 95% CI for the estimate $\hat{p}$ as well.

Sections 6.2.1 through 6.2.3 discuss the results of the SID experiment in three different ways—broken up by listeners, speakers, and then more directly by speech processing condition. When the data is broken up by listener, it is clear that each listener scored better than would be likely if he was randomly guessing. While there was a general decreasing trend in NTP, there were no statistically significant differences among sessions. Therefore, results broken up by session (or clip length) are not discussed. Some speakers are easier to recognize than others, and those not easily recognized are often confused with other speakers.

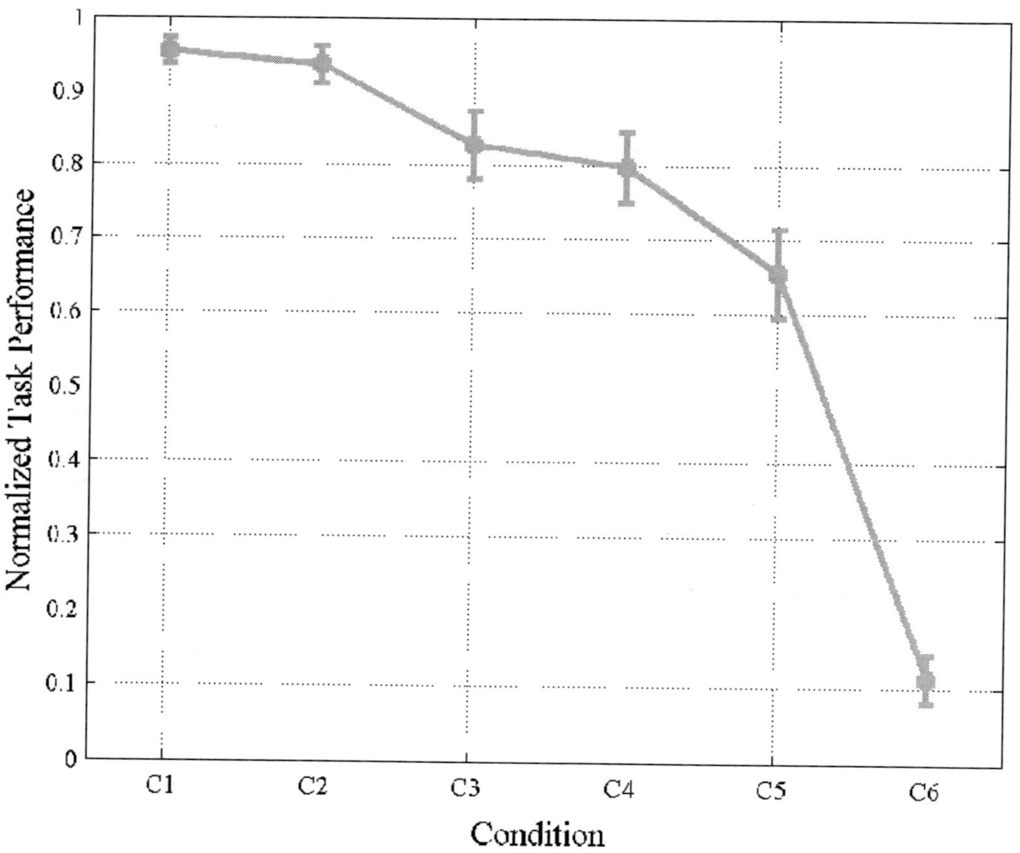

Figure 7. Intelligibility experiment: NTP mean and 95% CI for word intelligibility in sentence context.

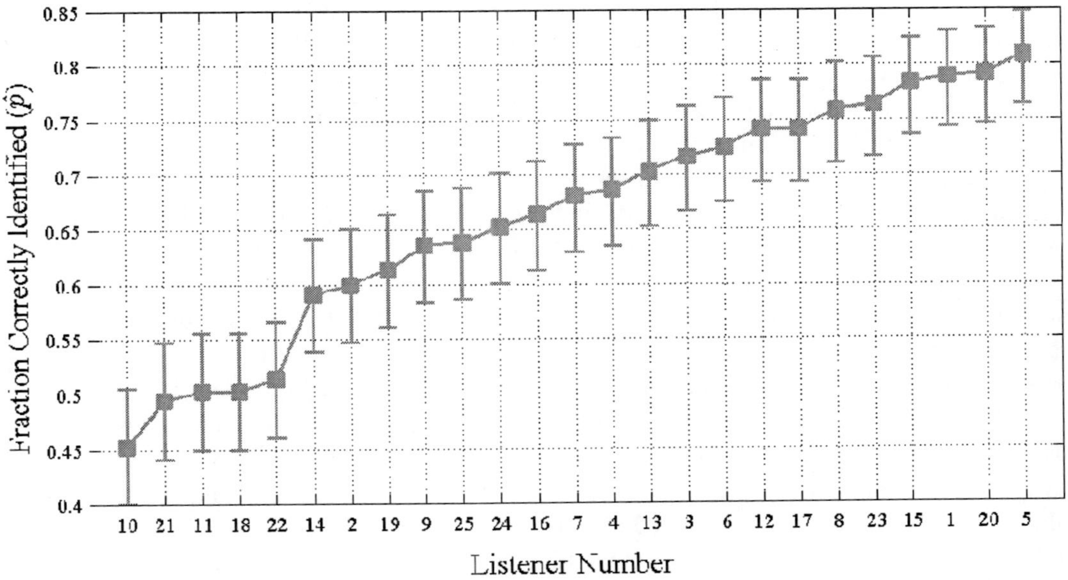

Figure 8. SID experiment: Mean fraction correctly identified by listener and 95% CI for each listener.

6.2.1. Listeners

Figure 8 shows the fraction of correct identifications and the associated 95% CI for the 25 listeners. Listeners are numbered and sorted by increasing fraction of correct identifications. The mean fraction of correct identifications over all listeners is 0.662, and 20 of the 25 listeners have overall correct identification fractions between 0.59 and 0.81. The people who utilized hearing aids have listener numbers 14 and 16. Listener 16 was also a non-native speaker. The listener who reported deafness in one ear is listener number 20. Figure 8 shows that none of these three listeners is an outlier. Thus, all three are retained in our data pool.

Out of the three listeners whose first language was not English, only one seemed to be at a disadvantage (listener 10). The other two non-native English speakers (listeners 13 and 16) placed close to the fraction-correct mean among all listeners. We elected to retain all three of these listeners in our data pool.

Not shown in Figure 8 is the experiment administrator, who received a great deal more training (more than 20 hours on speech distorted under all conditions). The experiment administrator was significantly more accurate with a 0.98 fraction of correct identifications. This is an indication that additional training can have a positive effect on SID performance. It's possible that the experiment administrator memorized to whom many of the recordings belonged, but it's also possible that over time he learned how each condition affected speech. The latter possibility may be relevant, as it is possible for listeners to become accustomed to adverse communication conditions. Once again, the experiment administrator's results were not included in the overall experiment results.

6.2.2. Speakers

Our selection of speakers had some interesting properties. The males had average pitches of 92, 105, and 111 Hz, and Male 3 had a slight Southern accent. The females had average pitches of 103, 104, and 107 Hz. Female 1 had a Midwestern accent, Female 2 had a Southern accent and Female 3 had a heavy Ecuadorian accent. The task of distinguishing among the three females is made easier (relative to the task of distinguishing among the three males) by very pronounced accents despite their small average pitch spread relative to that of the males.

Table 2 is the confusion matrix for the SID task for these six speakers averaged across all clips, conditions, and listeners. Each row is associated with one speaker, and each column is associated with the listener votes. "M" indicates male, "F" indicates female. Shaded cells indicate the fraction of correct SID, unshaded cells indicate fractions of confused SID. For example, the top left entry indicates that 67 percent of the clips from Male 1 were identified as coming from Male 1. The next entry to the right indicates that 22 percent of the clips from Male 1 were identified as coming from Male 2. Similarly, the next entry to the right indicates that 11 percent of the clips from Male 1 were identified as coming from Male 3.

The confusion matrix shows that Female 3 is easier to identify than Female 2, who in turn is easier to identify than Female 1 (correct identification fractions of 0.80, 0.74, and 0.65, respectively). Male 1 (with a correct identification fraction of 0.67) and Female 1 are close to the same difficulty, and Males 2 and 3 (fractions of 0.57 and 0.54, respectively) are both more difficult. The task of distinguishing among the males is difficult because Males 2 and 3 sound very similar (despite a slight Southern accent present in Male 3). In fact, the matrix shows that the greatest levels of confusion are between Males 2 and 3, though confusions between Male 1 and Male 2, and confusions between Female 1 and Female 2, are not far behind.

As Schmidt-Nielsen notes, listeners perform the SID task more efficiently with familiar, or distinctive speakers [6]. Our results are consistent with prior research. The two male speakers who had the smallest fraction of correct identifications were also often expressed as perceptually similar by listeners during the experiment. While the average pitch difference between the two easily confused male speakers is greater than the pitch spread among all female speakers, the female speakers were arguably more distinctive due to their regional accents. The speech processing conditions used in these experiments preserve pitch in the general sense, though the finer nuances of pitch trajectories will sometimes be lost.

Listeners received 15 minutes of training with these 6 unfamiliar speaker voices on average. Listeners were allowed to continue the training process as long as they wished. Many of the SID trials involved recordings in which the voice was greatly distorted. Thus, this amount of confusion is not unexpected. It is interesting to note that only Male 2 is ever perceived to be a female; all three females are confused for males, but only rarely.

Table 2. SID Experiment: Confusion Matrix Listener Selected Speaker.

	M1	M2	M3	F1	F2	F3
M1	0.670	0.220	0.110	0.000	0.000	0.000
M2	0.150	0.570	0.220	0.010	0.030	0.030
M3	0.120	0.340	0.540	0.000	0.000	0.000
F1	0.000	0.003	0.001	0.650	0.190	0.160
F2	0.000	0.004	0.001	0.170	0.740	0.080
F3	0.001	0.003	0.005	0.070	0.120	0.800

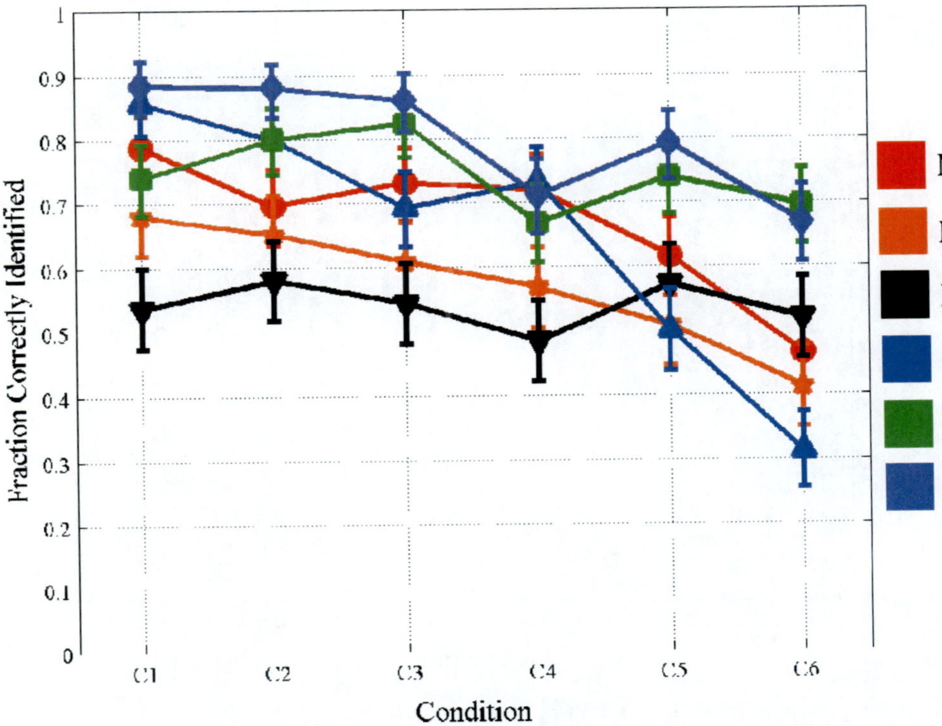

Figure 9. SID experiment: Fraction correctly identified and 95% CI by speaker and condition.

The difficulty of the SID task is broken down by speaker and by condition in Figure 9. Males 1, 2, and 3 are shown with red circles, orange stars, and dark gray downward-pointing triangles and lines, respectively. Females 1, 2, and 3 are shown with blue upward-pointing triangles, green squares, and purple diamonds and lines, respectively. With few exceptions, easier-to-identify speakers tend to be easier for all six conditions, and harder-to-identify speakers tend to be harder for all six conditions. The major exception is female 1 who is one of the easiest-to-identify speakers when heard over C1, C2, and C4, but is one of the hardest-to-identify speakers when heard over C5 and C6.

6.2.3.Speech Processing Conditions

A main goal of this work is to quantify how the listener SID performance is influenced by the six speech processing conditions. Each of the six conditions described in Table 1 was used for a total of 1,500 SID trials. For each condition, these 1,500 trials used the same 60 recorded speech files, and the same 25 listeners as well. This balance allows us to compare SID results for the six conditions directly, as Figure 10 shows.

For SID, the transformation from estimated probability of correct identification to NTP is

$$\text{NTP} = \frac{6}{5} \times \left(\hat{p} - \frac{1}{6} \right), \qquad \frac{1}{6} \le \hat{p} \le 1 \tag{2}$$

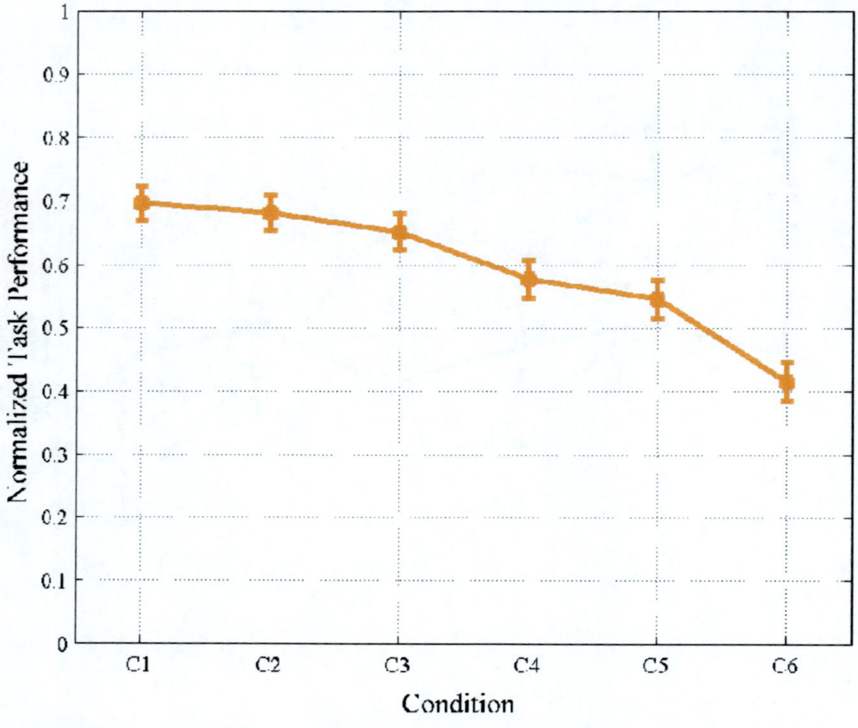

Figure 10. SID experiment: NTP mean and 95% CI by condition.

Because six responses are possible in this experiment, a listener making no effort and giving strictly random responses could have an average fraction of correct identifications of $\frac{1}{6}$. Thus, $\frac{1}{6}$ corresponds to no information from a listener, and (2) maps $\frac{1}{6}$ to an NTP value of 0. On the other hand, perfect SID corresponds to an NTP value of 1.

The transformation given in equation 2 is not defined for values of $\hat{p}$ that are less than $\frac{1}{6}$ (worse than guessing). While such values are theoretically possible, a meaningful interpretation is not obvious and would likely depend on additional factors. In this experiment no listeners exhibited performance worse than guessing. We expect that such results would be an extremely rare aberration and thus we do not consider the restriction in equation 2 to be a significant limitation.

Note that as we move from C1 to C6, the SID NTP drops steadily from 0.69 to 0.41. This is an NTP drop of 0.28.

6.3. Detection of Dramatized Urgency

For each trial in a detection experiment, three outcomes are possible: correct detection, false alarm (low urgency reported as high urgency), and miss (high urgency reported as low urgency). Given the binary nature of the data (correct or not correct), the maximum likelihood estimate for the probability of correct detection and the 95% CI for that estimate are calculated as given in (1) and [18]. As with SID, we report detection of DU results in terms of the NTP scale. In this case, that scale is defined by (3).

$$\text{NTP} = 2 \times \left(\hat{p} - \frac{1}{2} \right) \tag{3}$$

Because two responses are possible in this experiment, a listener making no effort and giving strictly random responses could have an average fraction of correct identifications of 21. (3) maps this to an NTP value of 0.

Figure 11 shows the mean NTP values and 95% CI for each of the six speech processing conditions, after averaging over all listeners and all messages for each condition. Figure 11 shows that as one progresses from C1 to C6, the NTP for detection of DU in messages drops from 0.76 to 0.58 (an NTP drop of 0.18). We also found that across the conditions, the false alarm rate tends to be lower than the miss rate. The false alarm rates generally fall into the range 0.05 to 0.10, while the miss rates generally span the range 0.10 to 0.35. In other words, detection-of-DU errors are less frequent when speakers are in the neutral state, and more frequent when speakers are in the DU state.

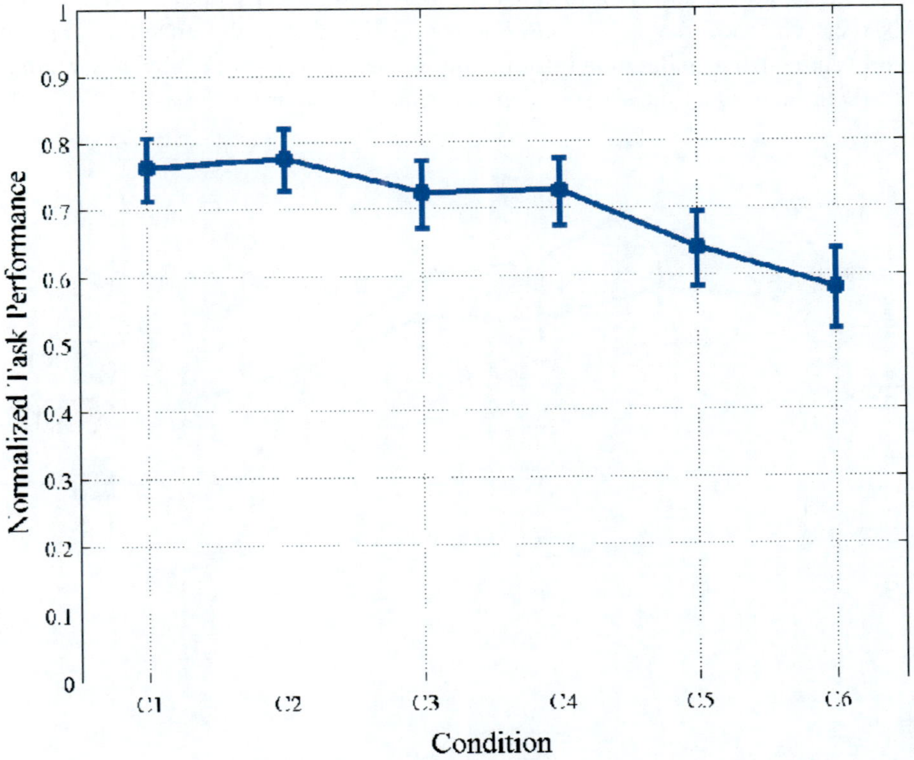

Figure 11. Dramatized Urgency Experiment: NTP mean and 95% CI for detection of dramatized urgency.

COMBINED RESULTS AND DISCUSSION

This section contains the results from 6.1, 6.2.3, and 6.3 combined. This allows one to compare the mean intelligibility, SID, and detection of DU results across the six speech processing conditions, consistent with the overall goal of this work. These combined results are given in Figure 12. The green line shows NTP mean and 95% CI for word intelligibility; the blue line shows mean and CI for detection of dramatized urgency; the yellow line shows mean and CI for SID. All three results generally decrease as one progresses from C1 to C6.

In the following we compare these decreases in intelligibility, DU, and SID NTP from C1 to C6 instead of comparing the raw NTP values of those three quantities for each speech processing condition. We believe that comparing these decreases leads to conclusions that are more likely to generalize beyond this specific set of experiments. It is clear that conclusions based on the raw NTP values do not generalize beyond this specific set of experiments because the raw NTP values can be very strongly influenced by experiment design.

If the SID experiment had used 24 speakers instead of 6 speakers, that task would have been much harder and the SID results curve in Figure 12 would undoubtedly be shifted downward dramatically, likely resulting in saturation with an NTP very close to 0 for all six speech processing conditions. Conversely, if the SID experiment had used just one female and one male speaker, that task would have been much easier and the SID results curve would have undoubtedly shifted upward dramatically, likely resulting in saturation with an NTP near 1 for all six speech processing conditions. Such dramatic shifts can saturate the NTP scale and this can prevent us from finding the information we seek in this work: how are intelligibility, DU, and SID influenced by the six speech processing conditions?

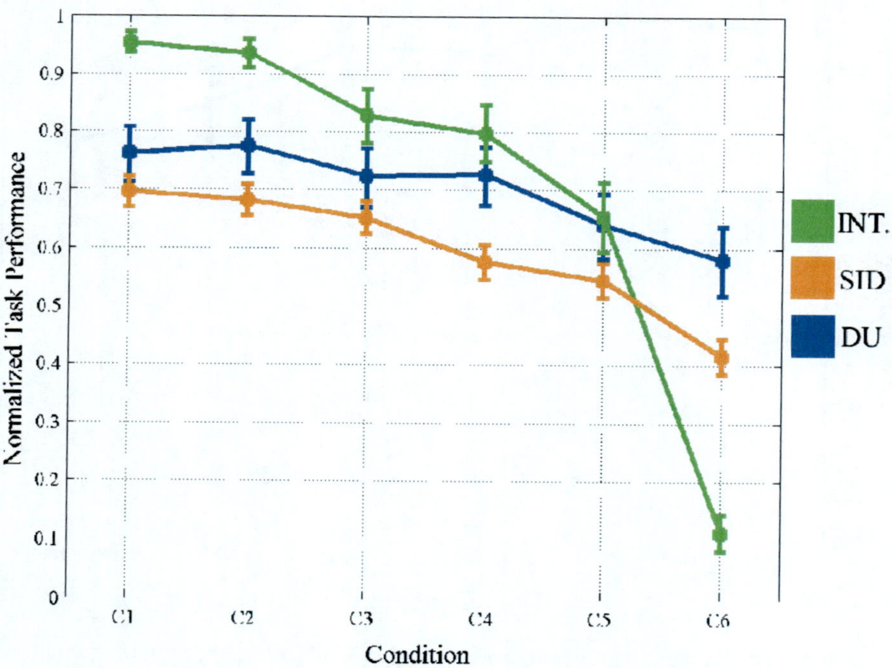

Figure 12. NTP mean and 95% CI for word intelligibility in a sentence context (green), detection of DU (blue), and SID (yellow).

Figure 12 shows that our experiment designs have been successful in that none of the three curves saturates at the top or bottom of the NTP scale. In light of the discussion above, statements about SID NTP relative to DU NTP for C3, for example, are specific to the parameters of these experiments (e.g., six talkers in SID, two levels of DU, and a host of other experiment parameters). On the other hand, an examination of the decreases in intelligibility, SID, and DU NTP as we progress from C1 to C6 leads to somewhat more general conclusions about the relative robustness of intelligibility, SID, and DU to the distortions induced by the six speech processing conditions. While these decreases in NTP may also be influenced by experiment design, we argue that they must be much less sensitive than the raw NTP values.

Figure 12 shows the intelligibility results decrease most abruptly (from 0.95 to 0.11 for a decrease of 0.84), and the detection of DU results decrease most gently (from 0.76 to 0.58 for a decrease of 0.18). The SID results show a decrease that is between the other two (from 0.69 to 0.41 for a decrease of 0.28). If we compare the NTP decrease for SID with the NTP decrease for intelligibility (0.28 compared with 0.84), we can conclude that the SID is 3.0 times (0. 84/0.28) more robust to the distortions created by the speech processing conditions than intelligibility is. Similarly, comparison of the NTP decrease for detection of DU with the NTP decrease for intelligibility (0.18 compared with 0.84) indicates that the detection of DU is 4.7 times (0.84/0.18) more robust to the distortions created by the speech processing conditions than intelligibility is.

These are the final relationships to be extracted from these experiments. They suggest that if a speech communication system is well represented by the speech processing conditions used in these experiments and its distortions are such that word intelligibility NTP has not decreased significantly (a very natural requirement for any useable system) then those distortions are unlikely to have caused SID and DU NTP to have decreased significantly.

Laboratory experiments like those described in this report are important because they provide a level of control over speaking, listening, and speech processing conditions that allows one to extract meaningful results. This would not be possible in a typical field environment. While laboratory experiments are essential to research progress, it is also true that the laboratory is often less realistic than the field environment.

One factor to consider in this regard is the consequence of various types and levels of background sounds at speaker and listener locations. It is clear that these background sounds can have negative effects, but they might also aid in SID (e.g., when it is known that Officer Roberts is at the coffee shop and Officer Smith is at a subway station, the corresponding background sounds could help with SID). They might also enable detection of urgency in speakers' voices. Examples of background sounds that may indicate a situation in which an officer could be exhibiting urgency in his voice may be screeching tires or even gunfire.

The relationship between dramatized urgency and the actual emotional signatures found in the voices of public safety officials is also of great interest. Dispatchers and officials who deal with urgent, catastrophic or tragic events on a routine basis may show less emotional variation in their voices than the general public would. However, it is possible that even officials with the calmest demeanor may find themselves in situations where they would exhibit urgency in their speech. Our results show that if this were to occur, the listener would likely be able to detect the sound of urgency. If intelligibility needed to be enhanced, officials could speak more slowly and clearly, but this could have a negative effect on SID, in that the speaker's normal cadence and speaking rhythm would be altered.

An additional issue centers on SID of familiar and unfamiliar speakers. Certainly years of professional association can cause voices and speaking styles to be very familiar, even under adverse conditions. This could lead to SID rates higher than those found in this laboratory study that uses unfamiliar speakers and a relatively short training or "acquaintance" process. On the other hand, many or most public safety officials communicate with far more than six other officials on a regular basis. This could make the SID task more difficult. How these two competing effects might balance out could only be determined through additional research efforts.

We designed the DU and SID experiments to have minimal complexity and yet provide useful basic DU and SID information. It is clear that one could design a large number of more intricate, more complex, or more specific experiments to explore the issues discussed in this section and a host of other associated issues. As experiments become more complex and specific, it becomes increasingly important to pay close attention to the balance between control and realism so that experiment results are of maximum utility and applicability.

REFERENCES

Cowie, R, Douglas-Cowie, E, Tsapatsoulis, N, Votsis, G, Kollias, S, Taylor, J & Fellenz, W. "Emotion recognition in human-computer interaction," *IEEE Signal Processing Magazine*, vol. 18, no. 1, pp. 32–80, Jan. 2001.

Bou-Ghazale S & Hansen, J. "A comparative study of traditional and newly proposed features for recognition of speech under stress," *IEEE Trans. Speech and Audio Processing*, vol. 8, no. 4, pp. 429–442, Jul. 2000.

Compton, AJ. "Effects of filtering and vocal duration upon the identification of speakers, aurally," *The Journal of the Acoustical Society of America*, vol. 35, no. 11, pp. 1748–1752, 1963.

Bricker PD & Pruzansky, S. "Effects of stimulus content and duration on talker identification," *The Journal of the Acoustical Society of America*, vol. 40, no. 6, pp. 144 1– 1449, 1966.

Uzdy, Z. "Human speaker recognition performance of LPC voice processors," *IEEE Transactions on Acoustics, Speech, and Signal Processing*, vol. 33, no. 3, pp. 752–753, Jun. 1985.

Schmidt-Nielsen A & Stern, KR. "Identification of known voices as a function of familiarity and narrow-band coding," *The Journal of the Acoustical Society of America*, vol. 77, no. 2, pp. 658–663, 1985.

Schmidt-Nielsen A & Stern, KR. "Recognition of previously unfamiliar speakers as a function of narrow-band processing and speaker selection," *The Journal of the Acoustical Society of America*, vol. 79, no. 4, pp. 1174–1177, 1986.

Schmidt-Nielsen, A. "A test of speaker recognition using human listeners," in *Proc. 1995 IEEE Workshop on Speech Coding for Telecommunications*, Annapolis, Maryland, Sep. 1995, pp. 15–16.

Schmidt-Nielsen A & Brock, DP. "Speaker recognizability testing for voice coders," in *Proc. 1996 IEEE International Conference on Acoustics, Speech, and Signal Processing*, Atlanta, Georgia, May 1996, vol. 2, pp. 1149–1152.

Schmidt-Nielsen A & Crystal, TH. "Human vs. machine speaker identification with telephone speech," in *Proc. 5th International Conference on Spoken Language Processing*, Sydney, Australia, Nov. 1998, vol. 2, pp. 221–224.

Quatieri, TF. *Discrete-Time Speech Signal Processing, Principles and Practice*, Chapter 14, Upper Saddle River, NJ: Prentice Hall, Inc., 2002.

Tactical Speaker Identification Database, Available at http://www.ldc.upenn.edu.

Steeneken H & Hansen, J. "Speech under stress conditions: Overview of the effect on speech production and on system performance," in *Proc. 1999 IEEE International Conference on Acoustics, Speech, and Signal Processing*, Mar. 1999, vol. 4, pp. 2079–2082.

Ruiz, R, Absil, E, Harmegnies, B, Legros, C & Poch, D. "Time-and spectrum-related variabilities in stressed speech under laboratory and real conditions," *Speech Communication*, vol. 20, no. 1-2, pp. 111–129, Nov. 1996.

Speech under Simulated and Actual Stress Database, Available at http://www.ldc.upenn.edu.

ITU-T Recommendation G. 191, Software tools for speech and audio coding standardization, Geneva, 2005.

ITU-T Recommendation P.8 10, Modulated noise reference unit (MNRU), Geneva, 1996.

Johnson, N, Kotz, S & Kemp, A. *Univariate Discrete Distributions*, New York: Wiley, second edition, 1992, p. 129.

<table>
<tr><td colspan="3">FORM NTIA-29 (4-80) U.S. DEPARTMENT OF COMMERCE
NAT'L. TELECOMMUNICATIONS AND INFORMATION ADMINISTRATION

BIBLIOGRAPHIC DATA SHEET</td></tr>
<tr><td>1. PUBLICATION NO.
TR-09-459</td><td>2. Government Accession No.</td><td>3. Recipient's Accession No.</td></tr>
<tr><td colspan="2" rowspan="2">4. TITLE AND SUBTITLE

Relationships Between Intelligibility, Speaker Identification, and the Detection of Dramatized Urgency</td><td>5. Publication Date
Nov. 2008</td></tr>
<tr><td>6. Performing Organization
NTIA/ITS.T</td></tr>
<tr><td colspan="2">7. AUTHOR(S)
Andrew A. Catellier and Stephen D. Voran</td><td rowspan="2">9. Project/Task/Work Unit No.

6513000-320</td></tr>
<tr><td colspan="2" rowspan="2">8. PERFORMING ORGANIZATION NAME AND ADDRESS
Institute for Telecommunication Sciences
National Telecommunications & Information Administration
U.S. Department of Commerce
325 Broadway
Boulder, CO 80305</td></tr>
<tr><td>10. Contract/Grant No.</td></tr>
<tr><td colspan="2">11. Sponsoring Organization Name and Address
National Institute of Standards and Technology/Office of Law Enforcement Standards
100 Bureau Drive, M/S 8102
Gaithersburg, MD 20899-8102</td><td>12. Type of Report and Period Covered</td></tr>
<tr><td colspan="3">14. SUPPLEMENTARY NOTES</td></tr>
<tr><td colspan="3">15. ABSTRACT
The systems used for public safety speech communications must be intelligible. It is also desirable that they transmit secondary information, such as the attributes of a speaker's voice. This secondary information can allow a user to identify the speaker and his or her emotional state. Testing speech communications systems for the delivery of intelligible speech is common. Testing for human perception of the delivery of this secondary information is less common, though some prior work has been done. Building on this prior work, we describe a set of controlled laboratory listening experiments. These experiments characterize the relationships between speech intelligibility, speaker identification, and the detection of dramatized urgency in a speaker's voice across a range of simulated speech processing conditions. The experiment results indicate that for the speech processing conditions considered here, detection of dramatized urgency is the most robust property, speaker identification is less robust, and speech intelligibility is the least robust.</td></tr>
<tr><td colspan="3">16. Key Words
human listening tests; intelligible speech; speaker identification; speaker stress detection; speaker urgency detection; speech transmission system; subjective speech quality tests</td></tr>
<tr><td rowspan="2">17. AVAILABILITY STATEMENT

☐ UNLIMITED.</td><td>18. Security Class. (This report)

Unclassified</td><td>20. Number of pages

35</td></tr>
<tr><td>19. Security Class. (This page)

Unclassified</td><td>21. Price:</td></tr>
</table>

NTIA Formal Publication Series

NTIA MONOGRAPH (MG)
A scholarly, professionally oriented publication dealing with state-of-the-art research or an authoritative treatment of a broad area. Expected to have long-lasting value.

NTIA SPECIAL PUBLICATION (SP)
Conference proceedings, bibliographies, selected speeches, course and instructional materials, directories, and major studies mandated by Congress.

NTIA REPORT (TR)
Important contributions to existing knowledge of less breadth than a monograph, such as results of completed projects and major activities. Subsets of this series include:

NTIA RESTRICTED REPORT (RR)
Contributions that are limited in distribution because of national security classification or Departmental constraints.

NTIA CONTRACTOR REPORT (CR)
Information generated under an NTIA contract or grant, written by the contractor, and considered an important contribution to existing knowledge.

JOINT NTIA/OTHER-AGENCY REPORT (JR)
This report receives both local NTIA and other agency review. Both agencies' logos and report series numbering appear on the cover.

NTIA SOFTWARE & DATA PRODUCTS (SD)
Software such as programs, test data, and sound/video files. This series can be used to transfer technology to U.S. industry.

NTIA HANDBOOK (HB)
Information pertaining to technical procedures, reference and data guides, and formal user's manuals that are expected to be pertinent for a long time.

NTIA TECHNICAL MEMORANDUM (TM)
Technical information typically of less breadth than an NTIA Report. The series includes data, preliminary project results, and information for a specific, limited audience.

End Notes

[125] The authors are with the Institute for Telecommunication Sciences, National Telecommunications and Information Administration, U.S. Department of Commerce, Boulder, CO 80305.

[2] The portraits shown in Figures 2-5 were used under Creative Commons license; attributions listed in the appendix.

In: Intelligibility Research and Communication Issues… ISBN: 978-1-61668-634-5
Editor: Samuel A. Fletcher

Chapter 3

EMERGENCY COMMUNICATIONS: VULNERABILITIES REMAIN AND LIMITED COLLABORATION AND MONITORING HAMPER FEDERAL EFFORTS

United States Government Accountability Office

WHY GAO DID THIS STUDY

Emergency communications breakdowns undermined response efforts during terrorist attacks in 2001 and Hurricane Katrina in 2005. In response, federal agencies like the Department of Homeland Security (DHS) and Federal Communications Commission (FCC) have increased efforts to enhance emergency communications. This requested report identifies (1) vulnerabilities, if any, to emergency communications systems; (2) federal assistance available or planned to first responders for addressing vulnerabilities or enhancing emergency communications; and (3) challenges, if any, with federal emergency communications efforts. GAO developed six catastrophic disaster case studies, reviewed agency documents, and interviewed public and private sector officials at the national, state, and local levels.

WHAT GAO RECOMMENDS

GAO recommends that DHS complete efforts to help implement the National Emergency Communications Plan; DHS and FCC establish a forum or other mechanism to collaborate on significant agency emergency communications efforts; and DHS leverage its expertise to help federal agencies develop emergency communications plans. DHS and FCC generally agreed with the recommendations. FCC raised concerns about the report's depth and scope. GAO clarified the scope and made other changes, as appropriate.

WHAT GAO FOUND

Continuity of communications, capacity, and interoperability are primary areas of vulnerability in first responder emergency communications in communities across the country. The destructive nature of catastrophic disasters can disrupt continuity of communications—the ability to maintain communications during and following a disaster. A volcanic mudflow at Mount Rainier, Washington, could destroy infrastructure supporting communications systems. Capacity—a communication system's ability to handle demand, provide coverage, and send different types of information—is also vulnerable in a catastrophic disaster. For example, blind spots, areas outside the range of communications systems, could inhibit response. Lastly, vulnerabilities involving interoperability—the ability to communicate across different organizations and jurisdictions as needed and authorized—remain due to technological and human factors.

Federal agencies provide a wide range of assistance intended to help first responders mitigate emergency communications vulnerabilities. GAO grouped available federal assistance into three categories: (1) new guidance and other significant federal efforts; (2) grants and funding; and (3) technical support and federal assets. DHS and other federal agencies have taken strategic steps to enhance emergency communications by issuing key documents like the National Emergency Communications Plan—the first strategic document for improving emergency communications nationwide. Numerous grants are available and are increasingly aligned with recently developed national and state plans. Federal agencies like DHS also offer technical support intended to help mitigate vulnerabilities through planning and on-the-scene assistance.

Limited collaboration and monitoring jeopardize federal emergency communications efforts, even as the federal government has taken strategic steps to assist first responders. Federal agencies have demonstrated limited use of some best practices that GAO previously reported as helpful for addressing issues like emergency communications. Delays in establishing the Emergency Communications Preparedness Center, which would help define common goals and mutually reinforcing strategies—two collaboration best practices—undermine the National Emergency Communications Plan's implementation. DHS and FCC have also not applied these practices in FCC's effort to promote a public safety network for emergency communications. Agency officials reported it was either too early or not the agency's responsibility to use these best practices in developing this network. DHS did not submit formal comments to FCC and FCC officials described its proposed network as separate from DHS emergency communications efforts. However, GAO found potential opportunities to align these agencies' efforts. Another collaboration best practice is leveraging resources, which DHS has done in providing emergency communications technical assistance and planning guidance. But efforts have focused on state and local jurisdictions and less on federal agencies, some of which lack formal emergency communications plans. Monitoring is also crucial in helping agencies meet goals.

ABBREVIATIONS

DHS	Department of Homeland Security
DOJ	Department of Justice
FCC	Federal Communications Commission
FBI	Federal Bureau of Investigation
FEMA	Federal Emergency Management Agency
MOU	memorandum of understanding
NOAA	National Oceanic and Atmospheric Administration
NCS	National Communications System
NTIA	National Telecommunications and Information Administration
SCIP	Statewide Communications Interoperability Plan
USGS	United States Geological Survey

June 26, 2009
The Honorable John F. Kerry
Chairman
Subcommittee on Communications, Technology, and the Internet
Committee on Commerce, Science & Transportation
United States Senate

Dear Mr. Chairman:

The September 11, 2001, terrorist attacks and response to Hurricane Katrina in 2005 exposed the severe consequences of breakdowns in emergency communications used by first responders. Failures in emergency communications resulted in numerous lost lives and exacerbated already challenging situations. These past events have increased focus on the need to enhance emergency communications to respond more effectively to future catastrophic disasters. Effective response to catastrophic disasters will require that first responders—law enforcement personnel, firefighters, and others first on the scene—have reliable communication systems, including supporting infrastructure, facilities, and staff. Such communication systems would enable first responders to communicate through voice, video, and other information seamlessly among themselves, various organizations, and different levels of government. Unless otherwise noted, when we refer to emergency communications systems, we mean those systems used by first responders. Since September 11, 2001, state and local jurisdictions, as well as the private sector, have invested billions of dollars to build and enhance existing communications systems.

Federal agencies have played and will continue to play an important role in supporting the further enhancement of emergency communications. The Department of Homeland Security (DHS) has led the development of guidance and equipment standards, as well as technological innovation. The Federal Emergency Management Agency (FEMA) within DHS has distributed grant funding, maintained and provided emergency communications assets, and developed assessment and planning tools for state and local jurisdictions. Other federal agency efforts are also underway. The Federal Communications Commission (FCC), an independent regulatory agency that oversees use of radio spectrum for non-federal entities, is currently pursuing the development of a nationwide, interoperable broadband network for public safety. Because catastrophic disasters can almost immediately overwhelm the response capabilities of state and local first responders, effective federal support before, during, and after such a disaster will be critical. That support may include providing communication assets, personnel, and support directly to state and local first responders. We have previously reported that best practices in collaboration and monitoring can aid federal agencies in addressing national, cross-cutting issues such as emergency communications.[1] In particular, we have found that given the importance of emergency communications and limited resources, it is critical that agencies find ways to work together to achieve effective and efficient outcomes.

In response to your request, this report focuses on issues related to emergency communications systems used by first responders in the aftermath of catastrophic disasters. Specifically, we identified and examined (1) vulnerabilities, if any, to emergency communications systems, (2) federal assistance available or planned to first responders for addressing any vulnerabilities or enhancing emergency communications, and (3) challenges, if any, with federal emergency communications efforts.

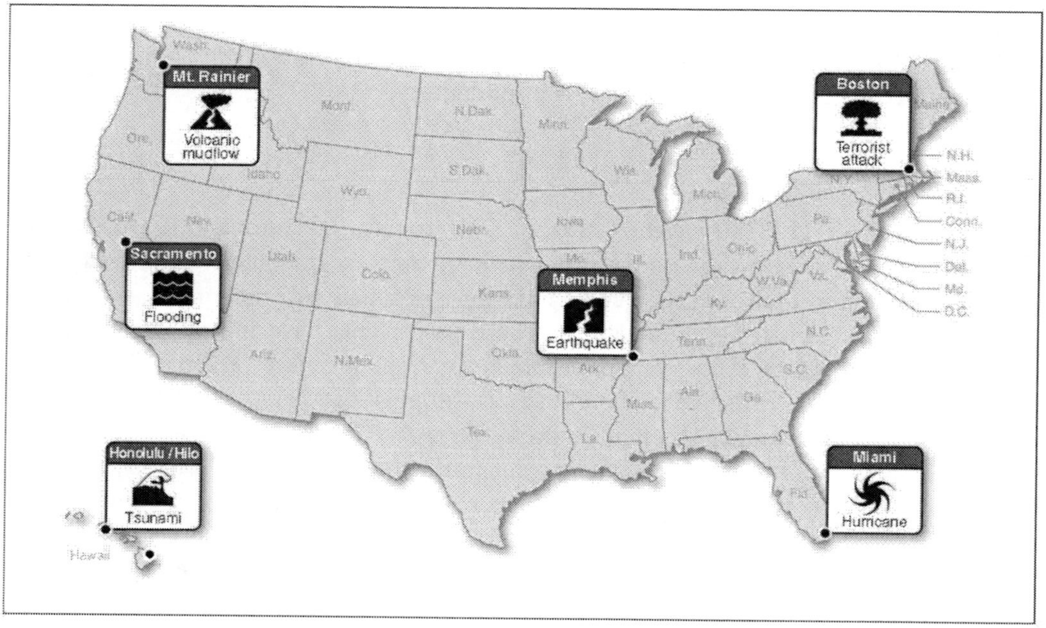

Sources: GAO and MapArt.

Figure 1. Emergency Communications Case Study Locations and Disaster Type.

To identify and examine vulnerabilities, if any, to existing emergency communications systems, we developed six case studies and subsequent analyses of varying catastrophic disaster scenarios both natural and manmade (see Figure 1). These case studies included a flood in northern California, a hurricane in southern Florida, a tsunami in Hawaii, a terrorist attack in Massachusetts, an earthquake in Tennessee, and a volcanic mudflow in the state of Washington. In selecting our case studies involving natural disasters, we conferred with subject matter experts from the National Oceanic and Atmospheric Administration (NOAA), United States Geological Survey (USGS), and other nongovernmental entities, as well as reviewed data on each respective location's natural hazards. We also considered factors such as the likelihood of occurrence, economic impacts, potential fatalities and injuries, and geographic diversity. For our case study involving a terrorist attack, we used scenario information produced by the Homeland Security Council[2] and selected a New England location to provide geographic diversity among our six case studies. We visited site locations for each of our six case studies and interviewed local and state emergency managers; police officers, firefighters, and other first responders; and regional federal officials to help identify emergency communications vulnerabilities. We also conducted a literature review of our prior products and other federal agency reports on emergency communications to analyze and ascertain common vulnerabilities.

To identify and examine federal assistance available to first responders for emergency communications, we interviewed officials and reviewed program documents from a variety of federal agencies with responsibility for emergency communications efforts available or planned, such as DHS, FCC, and the Department of Justice (DOJ). During our case study work, state and local first responders, as well as federal officials, also provided information on federal efforts that we report on. To identify and examine any challenges in the federal approach to supporting emergency communications, we consulted our related past work on emergency communications, interagency collaboration, and federal government program management and performance. We analyzed key federal agency documents, such as DHS's National Emergency Communications Plan and FCC's notices for proposed rulemaking for an interoperable nationwide broadband public safety network to determine the extent of interagency collaboration and monitoring in some significant federal efforts. We interviewed federal agency officials to determine what steps had been taken by their respective agencies to collaborate and monitor these efforts. We also interviewed state and local first responders, professional and trade group representatives, and officials in the telecommunications industry to obtain their perspectives on significant federal efforts.

We conducted this performance audit work from February 2008 to May 2009 in accordance with generally accepted government auditing standards. Those standards require that we plan and perform the audit to obtain sufficient and appropriate evidence to provide a reasonable basis for our findings and conclusions based on our audit objectives. We believe that the evidence obtained provides a reasonable basis for our findings and conclusions based on our audit objectives.

Results in Brief

Continuity of communications, capacity, and interoperability are the primary areas of vulnerability in emergency communications that persist in communities across the country, based on interviews with state and local first responders in our six case studies and others, as well as a literature review, including our prior work. DHS and FCC have described similar vulnerabilities. First responders also noted that communications vulnerabilities extend beyond our case study locations and disaster scenarios.

- Through powerful effects, such as high winds and ground shaking, potential catastrophic disasters can disrupt continuity of communications—the ability to maintain communications during and following a disaster—by destroying infrastructure supporting communications systems. For example, a volcanic mudflow at Mount Rainier National Park in Washington state could destroy cable supporting phone communications. In addition, disasters may limit continuity of communications by damaging communications facilities and stranding first responders. For example, a major earthquake in Tennessee could damage roads and bridges, stranding Memphis first responders across the Mississippi River.

- Limitations in system capacity—a communication system's ability to handle demand, provide coverage, and send different types of information—could inhibit response. Spikes in demand following a disaster can cause communications systems to crash and system outages place additional demands on remaining systems. In addition, areas outside of the range of local communications systems can limit response efforts by creating "blind spots" in coverage, such as those found in Hawaii's mountainous terrain. Furthermore, some equipment may lack the capacity to send photographs and video, reducing first responders' situational awareness.
- We have previously reported on vulnerabilities involving interoperability— the ability to communicate across different organizations and jurisdictions as needed and authorized—and first responders we interviewed identified technological and human factors that continue to limit interoperability. Jurisdictions use various, and at times incompatible, communications systems. For example, some fire departments have hesitated to use digital radio systems, which could create incompatibility with other first responder systems, such as law enforcement. The fast-changing nature of technology compounds the difficulty of fostering and maintaining interoperability. Human factors can also limit interoperability, such as the increasingly critical need to have staff trained to coordinate with a growing number of jurisdictions.

Federal agencies provide a wide range of assistance intended to help first responders mitigate emergency communications vulnerabilities, which we grouped in three categories: (1) new guidance and other significant federal efforts, (2) grants and funding, and (3) technical support and federal assets. Recently, DHS and other federal agencies have taken significant and strategic steps to enhance emergency communications by issuing the National Emergency Communications Plan in July 2008, the first strategic document focused exclusively on improving emergency communications nationwide. Other recent federal efforts underway include completing a memorandum of understanding (MOU) to establish the Emergency Communications Preparedness Center—to be jointly operated by a number of federal agencies such as DHS, FCC, and the Department of Commerce, as the focal point and clearinghouse for implementing federal interoperability efforts—and establishing multiple DHS and FCC stakeholder groups to formulate recommendations for improving emergency communications based on lessons learned from previous disasters. The second category of assistance includes a wide range of grants and funding, some of which are increasingly aligned with recently developed national and state plans. Finally, federal agencies such as DHS offer technical support and assets intended to help mitigate emergency communications vulnerabilities, both through advanced planning and onthe-scene assistance. For example, DHS has developed programs such as the Interoperable Communications Technical Assistance Program, providing support to first responders for planning and technical issues to be considered when developing interoperable communications.

Limited collaboration and monitoring jeopardize progress in emergency communications, even as the federal government has taken significant and strategic steps to assist first responders. Federal agencies have demonstrated limited application of some collaboration best practices that we have previously reported as helping address issues like emergency communications, which are national in scope and cross agency jurisdictions. For example, delays in establishing the Emergency Communications Preparedness Center, which would help define common goals and mutually reinforcing strategies—two collaboration best practices—undermine the implementation of the National Emergency Communications Plan, which relies heavily on participation from multiple agencies. Additionally, DHS and FCC have not established a common vision or mutually reinforcing strategies for a nationwide broadband public safety policy, although both agencies play key roles in such a development—DHS as the agency responsible for developing and overseeing the National Emergency Communications Plan and FCC as the agency charged with overseeing spectrum for non-federal entities. Although FCC has for the last several years been engaged in an effort to promote a nationwide interoperable broadband network for public safety ("700 MHz Public/Private Partnership"), there has been limited coordination with DHS. According to officials from DHS and FCC, it was either too early or not the agency's responsibility to undertake these best practices for this effort. DHS did not submit formal comments to FCC during its most recent 700 MHz Public/Private Partnership comment period. FCC officials described the 700 MHz Public/Private Partnership and the National Emergency Communications Plan as two separate, but parallel efforts. However, based on our analysis, we found potential opportunities to align DHS and FCC emergency communications efforts. Another collaboration best practice is leveraging resources. While DHS has leveraged its expertise in emergency communications planning to provide technical assistance and guidance, these efforts have focused on state and local jurisdictions, and less so on other federal agencies, some of which do not have formal emergency communications plans. We have also previously reported that monitoring and evaluating efforts are crucial elements to achieving agency goals. Although DHS and FCC have various ways of examining stakeholder group recommendations, neither agency systematically monitors or evaluates recommendations from agency- assembled stakeholder groups or the agency's response either, potentially limiting the groups' relevance and value in addressing vulnerabilities.

We make four recommendations in this report to improve federal agencies' collaboration and monitoring in efforts related to emergency communications. To help foster implementation of the National Emergency Communications Plan, we are recommending that the Secretary of Homeland Security work to complete a memorandum to establish the Emergency Communications Preparedness Center. To help ensure that significant federal efforts are collaborative, we are recommending that the Secretary of Homeland Security and the Chair of FCC establish a forum, or other mechanism, to better collaborate to identify and discuss challenges, opportunities, and potential ways to better align their emergency communications efforts, such as the National Emergency Communications Plan and the 700 MHz Public/Private Partnership. To help ensure that federal agencies are well-positioned to support state and local first responders in a disaster, we are recommending that the Secretary of Homeland Security provide guidance and technical assistance to federal agencies in developing formal emergency communications plans. Finally, to enhance the value of DHS and FCC stakeholder group recommendations, we are recommending that the Secretary of

Homeland Security and the Chair of FCC systematically track, assess, and respond to stakeholder groups' recommendations.

We provided a draft of this report, for official review and comment, to DHS, FCC, Commerce, Interior, and DOJ. DHS generally agreed with our recommendations and provided comments that are discussed near the end of this letter. DHS's comments are contained in appendix V. FCC provided comments via e-mail and agreed with our recommendations, but raised concerns that related to the depth and scope of our analysis, such as stating that the report relies heavily on anecdotes and opinion. We made changes to clarify the scope of our work, but remain confident about our findings and conclusions. We discuss FCC's comments in detail near the end of this letter. The comments from Commerce and Interior are discussed near the end of this letter and contained in appendixes VI and VII, respectively. Interior commented that the report could have been improved by incorporating Interior or federal interoperability collaboration efforts in regards to emergency response capabilities. DOJ did not comment on the report. DHS, FCC, Commerce, and Interior also provided technical comments that we incorporated, where appropriate.

BACKGROUND

Many regions of the country face hazards from natural and man-made disasters, some of which could prove catastrophic (see figure 2). Unlike most typical disasters, catastrophic incidents can yield extraordinary levels of mass casualties, damage, or disruption, immediately overwhelming the response capacities of state and local resources, and requiring outside action and support from the federal government and other entities.[3] Some catastrophic disasters, such as large-scale hurricanes, may be detected or forecast well before they impact population centers, though their intensity and path can change significantly and quickly. Other catastrophic incidents, such as earthquakes and terrorist attacks, can occur with little or no notice. DHS has encouraged an all-hazards approach to disaster planning, to ensure that communities consider all threats faced, both natural and man-made, in the planning process. An all-hazards approach accounts for vulnerabilities, such as damage to infrastructure, that occur in various types of disasters in locations across the country. Some types of disasters, such as hurricanes, are more likely to occur in certain areas of the country, but many regions face hazards from one or multiple types of disaster. The goal of disaster preparedness and response is to prevent where possible, prepare for, or mitigate, and respond to disasters of any size or cause with effective actions at all levels of government that minimize the loss of life and property and set the stage for a quick recovery.[4]

First responders play a critical role in disaster preparedness and recovery, assisting in the response to emergency events, including catastrophic disasters. Typically, first responders include law enforcement, firefighters, emergency medical personnel, and others who are among the first on the scene of an emergency. However, since the terrorist attacks of September 11, 2001, the definition of first responder has grown to include other organizations, such as public health and hospital personnel, which may not be on the scene, but are essential in supporting effective response and recovery operations. Depending on the nature and location of a catastrophic event, responders on the scene may also include federal agencies directing all or a portion of the federal disaster response or assisting state and local first responders in their response efforts. For example, the Federal Bureau of Investigation (FBI) would participate in the response to a terrorist attack, based on its mission to protect and defend the United States against terrorist threats.

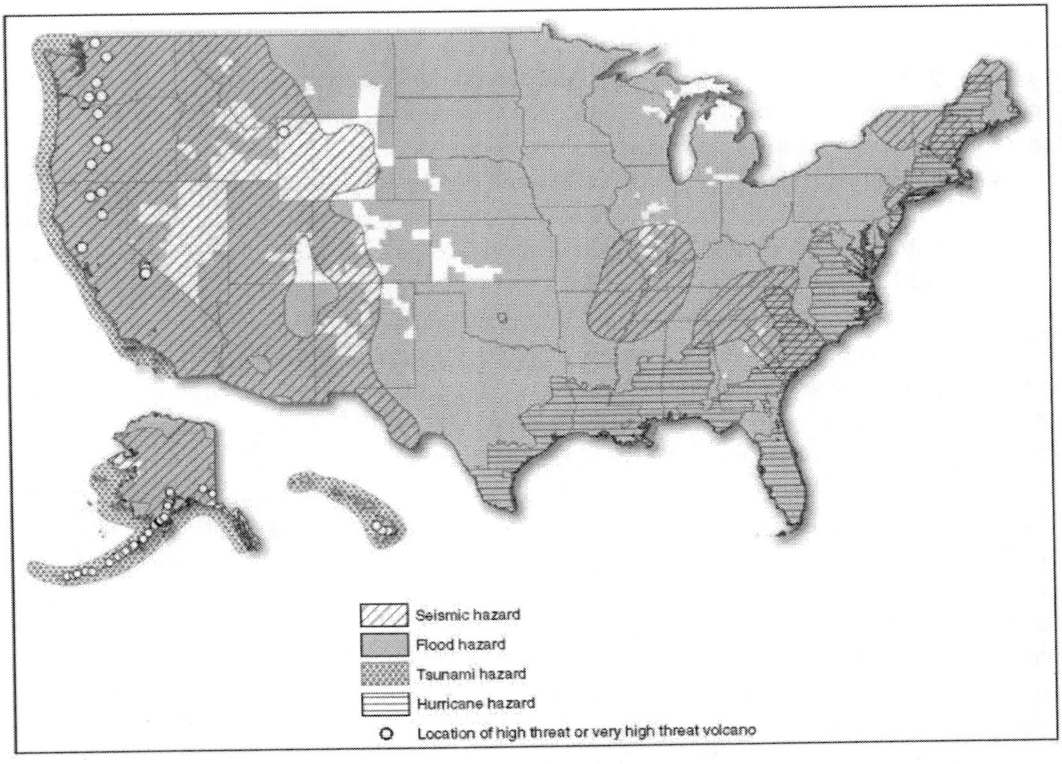

Sources: GAO analysis of Federal Emergency Management Agency (FEMA), National Oceanic and Atmospheric Administration (NOAA), and United States Geological Survey (USGS) data; Map Resources (map).

Note: Figure 2 depicts only those natural disasters included as part of our case study work and omits other disaster types, such as tornadoes. We derived hurricane hazards in the figure from hurricane strike data from 1980 through 2007. NOAA officials noted that the impact of hurricanes can be felt along the U.S. Gulf and Atlantic Coasts from Texas to Maine and extend inland for hundreds of miles.

Figure 2. Examples of Natural Disaster Hazards in the United States.

Communications systems serve as the backbone for first responders in gathering and sharing information, coordinating response, and requesting additional resources and assistance from neighboring jurisdictions and/or the federal government. Effective communications are vital to first responders' ability to respond and ensure the safety of both their personnel and the public. First responders cooperate to rescue victims, oftentimes relying on several different communications systems to do so. Voice, data, and video technology, if available, can be used to share information seamlessly between first responders, other various organizations, and different levels of government.

Recent catastrophic events have underscored the importance of emergency communications. For example, the 9/11 Commission concluded that the large number of deaths among firefighters during the collapse of the World Trade Center was partly attributable to a communications failure.[5] Following the September 11, 2001, terrorist attacks and Hurricane Katrina in 2005, Congress expanded a number of federal agencies' roles and responsibilities related to emergency communications. The Homeland Security Act of 2002 established DHS and required the agency, among other things, to build a comprehensive national incident management system comprising all levels of government and to consolidate existing federal government emergency response plans into a single, coordinated national response plan.[6] Hurricane Katrina highlighted additional communications challenges and demonstrated the need to improve emergency communications leadership at all levels of government in order to better respond to a catastrophic disaster.

More recent legislation has directed DHS and FEMA to take on certain roles and actions related to emergency response and emergency communications. To address many of the challenges highlighted by the Hurricane Katrina response, the Post-Katrina Emergency Management Reform Act of October 2006 (Post-Katrina Act) was enacted, and established within DHS, the Office of Emergency Communications to help develop, implement, and coordinate interoperable and operable communications for the emergency response community at all levels of government.[7] The Office of Emergency Communications also oversees other DHS efforts, including elements of the SAFECOM program,[8] and the development of the National Emergency Communications Plan and other key documents intended to create an overarching strategy to address emergency communications shortfalls. The Post-Katrina Act also charged FEMA with the primary responsibility for coordinating and implementing key aspects of federal emergency preparedness and response, including grants management. As required by the act, FEMA is to lead and support the nation in a risk-based, comprehensive emergency management system of preparedness, protection, response, recovery, and mitigation. FEMA leads the integration of tactical federal emergency communications during disasters and often deploys personnel or equipment to the scene of a disaster to manage the federal response.

Other federal agencies also have a role in emergency communications and disaster response. For example, in September 2006, FCC established its Public Safety and Homeland Security Bureau, which is responsible for developing, recommending, and administering the agency's policies pertaining to public safety communications issues.[9] The bureau submits annual reports to the FCC Chairman and Commissioners and hosts quarterly summits on various topics relevant to the public safety community. In addition, the bureau has established a clearinghouse to collect, evaluate, and disseminate public safety information. FCC also manages the use of the radio-frequency spectrum by non-federal entities, such as commercial enterprises, state and local governments, and public safety organizations. Radio spectrum is a fixed, limited resource, which government and nongovernmental entities share for commercial and public safety communications.[10] In 1993, legislation authorized FCC to use competitive bidding—or auctions—to assign spectrum licenses to commercial users.[11] For the last several years, FCC has pursued a new Public/Private Partnership (the 700 MHz Public/Private Partnership) in a proceeding involving commercial and public safety spectrum in the 700 MHz Band, which was occupied by television broadcasters.[12] As part of the digital television transition, this spectrum was to be cleared and made available for public safety and commercial services in June 2009.[13] See figure 3.[14]

Another agency with a role in emergency communications is the Department of Commerce's National Telecommunications and Information Administration (NTIA), which is responsible for managing spectrum used by the federal government.[15] Officials from NTIA and other agencies also serve on a number of interagency committees to coordinate their activities on a standing and disaster-activated basis. DHS's National Communications System (NC S) coordinates the emergency support function for communications, which involves, among other things, oversight of communications within the federal incident management and response structures. Interior has also been an active joint federal partner through the National Interagency Fire Center, which has provided search and rescue capabilities, as well as deploying, operating, and managing communications systems during recent disasters.

With a mission to ensure public safety against foreign and domestic threats, DOJ has also worked with other federal agencies, such as DHS and the Department of the Treasury, to improve disaster response. For example, in 2001, DOJ initiated an effort to provide secure, seamless, and interoperable wireless communications for federal agents and officers engaged in law enforcement, homeland defense, and disaster response.[16]

Multiple federal agencies have a role in disaster preparedness and response, and there are several best practices agencies can employ to help overcome the barriers to successful inter-agency collaboration. We have previously reported on collaboration best practices, which are useful in addressing issues that are national in scope and cross agency jurisdictions, such as emergency communications.[17] For the purposes of our report, we focus on the three best practices described in table 1. Prior GAO work has also shown that monitoring and evaluating agency actions and progress can help key decision-makers obtain feedback for improving both policy and operational effectiveness.[18]

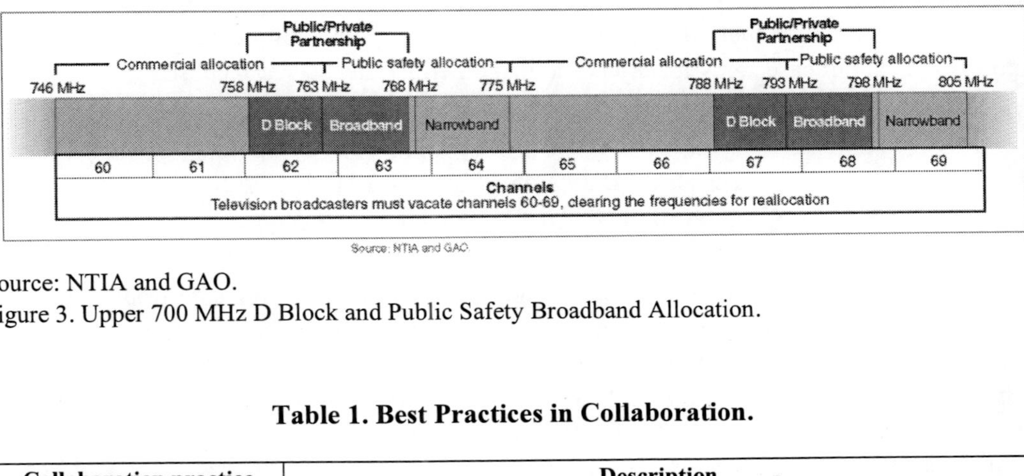

Source: NTIA and GAO.
Figure 3. Upper 700 MHz D Block and Public Safety Broadband Allocation.

Table 1. Best Practices in Collaboration.

Collaboration practice	Description
Define and articulate a common outcome	Collaboration requires agency staff working across agency lines to define and articulate the common federal outcome or purpose they are seeking to achieve that is consistent with their respectiveagency goals and mission.
Establish mutually reinforcing or joint strategies	To achieve a common outcome, collaborating agencies need to establish strategies that work in concert with those of their partners or are joint in nature. Such strategies help in aligning the partner agencies' activities, core processes, and resources to accomplish the common outcome.
Identify and address needs by leveraging resources	Collaborating agencies bring different levels of resources to the effort. Collaborating agencies can look for opportunities to address resource needs by leveraging each other's resources, thus obtaining additional benefits that would not be available if they were working separately.

Source: GAO.

To improve emergency preparedness, states, regions, and local jurisdictions have also invested billions to build dedicated networks and acquire technology, lease or subscribe to private carrier services for primary or backup systems, and to maintain and test existing communications systems. Similarly, private stakeholders, such as telecommunications companies and equipment manufacturers, have invested heavily to develop innovative technological solutions and expand or strengthen their networks for emergency responders and commercial use. Private stakeholders develop proprietary technology and networks that first responder agencies may buy, lease, or subscribe to by paying service charges (see app. III for an overview of some of the various technologies that first responders use).

Continuity of Communications, Capacity, and Interoperability Are Primary Areas Where Emergency Communications Remain Vulnerable

Continuity of communications, capacity, and interoperability are the primary areas of vulnerability in emergency communications that persist in communities across the country. We identified these vulnerabilities in interviews with state and local first responders in each of our six case studies and others, as well as a review of emergency communications literature, which include our prior work. DHS and FCC have identified similar vulnerabilities in recent work, including continuity of communications and interoperability.[19] First responders also noted that identified communications vulnerabilities extend beyond the communities in our case study locations and that other disaster scenarios pose similar hazards.

Catastrophic Disasters Threaten Continuity of Communications

Destructive forces, such as high winds and ground shaking, during catastrophic disasters can disrupt first responders' continuity of communications—the ability to maintain communications during and following a disaster—in a number of ways. Using the scenario of a major earthquake in a city, figure 4 depicts how damage to supporting infrastructure and communications facilities, as well as stranding first responders, may threaten continuity of communications.[20]

Damage to Supporting Infrastructure

Communications systems used by first responders, such as landline phone systems and certain radio systems, cannot function without phone cables, radio towers, and other supporting infrastructure. For example, Hurricane Katrina's high winds and flooding destroyed emergency communications infrastructure in Louisiana and Mississippi, disrupting continuity of communications in several states and inhibiting the response. Potential, future catastrophic disasters pose similar hazards, such as a lahar—a volcanic mudflow—in Mount

Rainier National Park in Washington state.[21] National Park Service officials stationed in the park said that the park's telephone system relies on a privately-owned phone cable, which is old and exposed in many locations (see figure 5). According to park officials, keeping the cable operational is a constant challenge even under normal circumstances. In the event of a lahar at Mount Rainier, fast-moving mud and debris could destroy the cable and disrupt the park's phone system (see app. II for more information on the hazards associated with our six case studies).

Damage to Communications Facilities

Emergency operations centers[22] and other facilities serve as command posts from which first responders coordinate and launch a response. Yet a disaster may damage or destroy these facilities, rendering them useless. First responders in Jackson, Tennessee, described intense tornadoes hitting and damaging two emergency operations centers in 2003, which then inhibited the response. Responders in Jackson and Memphis, Tennessee, also said that some of their facilities were vulnerable to future earthquakes generated in the New Madrid seismic zone.[23]

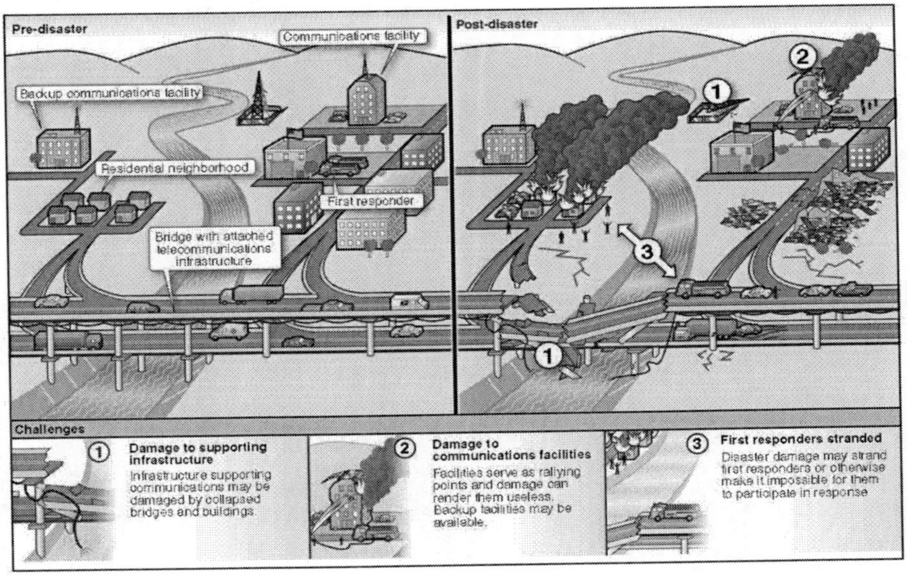

Source: GAO.

Figure 4. Vulnerabilities Involving Continuity of Communications in an Earthquake Scenario.

Local facilities were not constructed to withstand seismic shaking, and some are located on thick sediment, which can amplify seismic shaking. Even if a facility experiences little direct damage, the disaster may down power lines to the facility, which some communications systems need to function. Officials at the National Public Safety Telecommunications Council described maintaining power as the most basic vulnerability facing emergency communications after a disaster. Officials at the Central United States

Earthquake Consortium [24] noted that facilities in Tennessee and neighboring states have backup power generators. However, some fuel tanks po wering the generators are not prope rly secured and may otherwise be vulnerable to seismic shaking (see figure 6).

Source: GAO.

Figure 5. Exposed, Hanging Cable at Mount Rainier National Park.

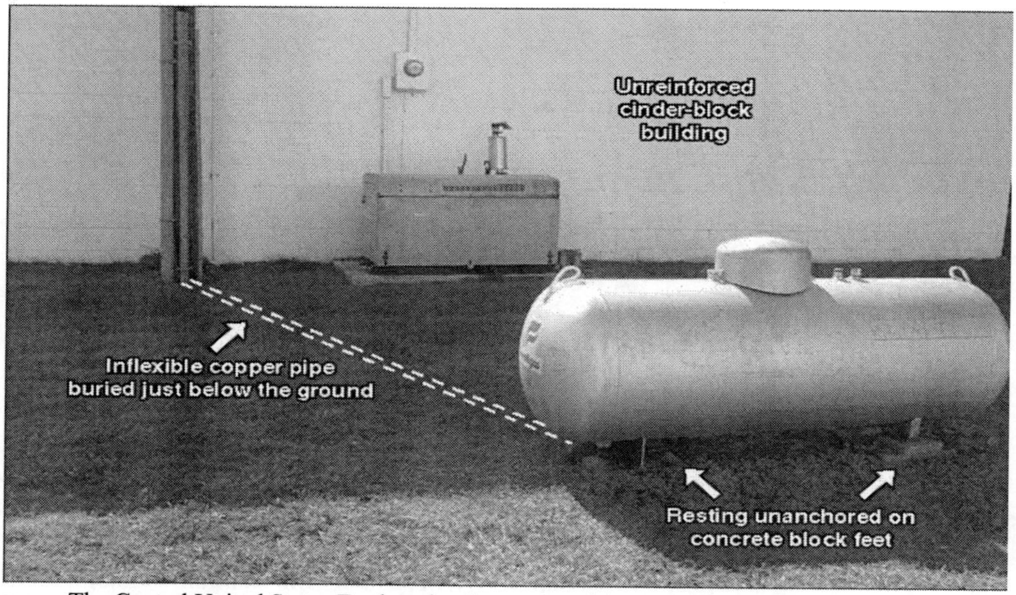

Sources: The Central United States Earthquake Consortium (photograph); and GAO.

Figure 6. Vulnerable Fuel Tank in the New Madrid Seismic Zone.

To help maintain continuity of communications, some jurisdictions have designed facilities to withstand damages expected from future disasters. For example, Miami-Dade County officials described mitigating potential hurricane wind damage at a county communications facility by adding a sloped roof. First responders in Memphis, Tennessee reported that some newer facilities had been built to supposedly resist seismic shaking. Tennessee and California first responders identified backup locations that they would move to if their facilities were damaged, but this move would take time before they would regain full communications capabilities.

Stranded First Responders

Disasters may strand first responders or otherwise make it impossible for them to participate in a response. Following Hurricane Katrina, many state and local first responders were incapacitated and flooding blocked access. This inhibited response by preventing the establishment of a command structure for the response, reducing communications and awareness of the situation following Hurricane Katrina's landfall. Memphis first responders expressed concerns that a future, major earthquake in the New Madrid seismic zone will damage bridges and strand some first responders across the Mississippi River in Arkansas. With the Mississippi river bisecting the region, bridges are some of the most important and seismically vulnerable piece of the transportation network. A majority of the bridges were designed with little or no seismic consideration. Law enforcement and fire department officials in Tennessee, Florida, and Washington state conveyed concerns about the ability of their staff to report after a major disaster. Damage to first responders' property, as well as personal injury to themselves or family, can also prevent participation in response. To address such difficulties and maintain continuity of communications, Miami-Dade County has taken steps to enable responders' families to shelter in local facilities and to help repair first responder property damage.

Limited System Capacity Hinders First Responders' Communications

A number of capacity issues can hamper emergency communications systems used in disaster response. For our work, we use the term "capacity" to refer to a communication system's ability to handle demand, provide coverage, and send different types of information (i.e., voice and data). Using the scenario of a lahar hitting a small town, figure 7 depicts how capacity may be threatened by system crashes due to demand, the existence of blind spots, and an inability to use non-voice communications.

System Crashes due to Demand

Some communications systems used by first responders may lack the capacity to prevent system crashes due to spikes in demand, which can follow disasters. Telecommunications company officials reported that their systems are not designed to handle everyone in a region calling simultaneously. Past disasters, such as the terrorist attacks in 2001 and Hurricane Katrina in 2005, created excessive demand, which caused communications system to crash. System outages can also place additional demands on remaining systems. More recently, officials in the California Governor's Office of Emergency Services reported that over 5 million calls followed a moderate earthquake and disrupted communications for a short time. FCC has reported that first responders enjoy communications capabilities that are more robust than those provided by the private sector; yet, communications also rely on the functioning of the 85 percent of the nation's critical communications infrastructure that the private sector controls. Boston Fire Department officials told us that they anticipate reduced communications capabilities following a disaster due to system crashes. Jurisdictions are working to increase capacity on public and private sector communications systems and related infrastructure. For example, some jurisdictions are building new fiber optic networks. In addition, some telecommunications companies offer jurisdictions services for additional system capacity in a disaster, such as "cell on wheels" and "cell phone on light trucks," to restore communications.[25]

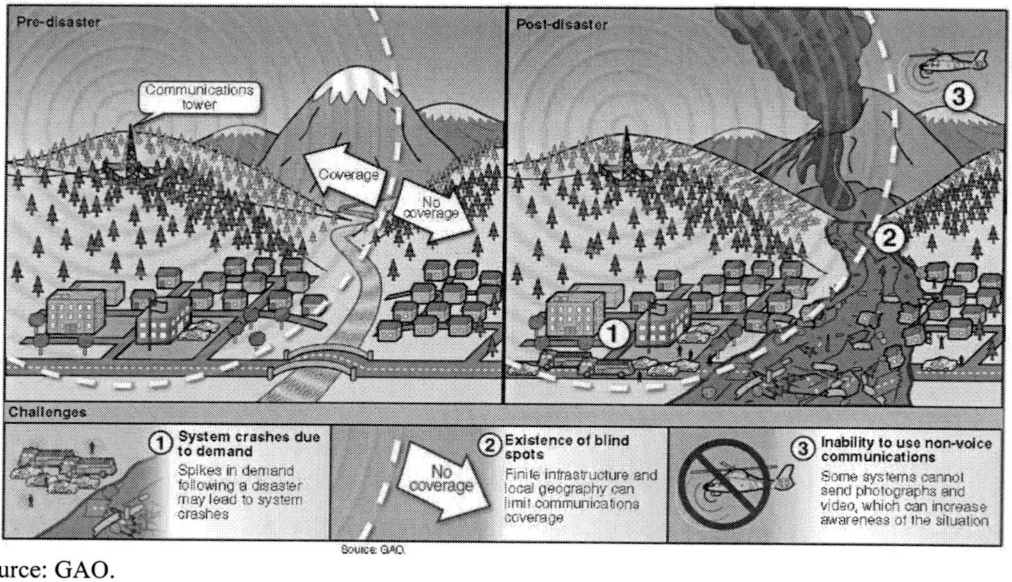

Source: GAO.

Figure 7. Vulnerabilities Involving Capacity Limitations in a Lahar—Volcanic Mudflow—Scenario.

Existence of Blind Spots

Communications system "blind spots"—that is, areas that lie outside the range of local communications systems—exist for a number of reasons. Some communications systems have finite infrastructure, such as radio systems with a limited number of towers and effective transmission range. In addition, local geography can create blind spots as elevation changes or high-rise buildings interfere with radio signals. We observed instances of system blind spots in our case studies. According to Hawaii first responders, mountainous terrain has created blind spots for some communities near the water, which could inhibit emergency communications and response during a tsunami.[26] Law enforcement officials in one of our other case study locations also reported that some local tunnels are blind spots for certain emergency communications systems in the area. Jurisdictions are addressing blind spots by investing in mobile communications vehicles. In the event of an existing blind spot, or damage from a disaster creating new ones, these vehicles can plug gaps in emergency communications coverage by establishing a mobile communications network at or near the scene of an incident. Vehicles are equipped with cellular and satellite phone and fax capabilities, an on-board computer network, printers and satellite, internet access, video teleconferencing, recording, and broadcast/satellite television (see figure 8). However, such assets are not a cure all for blind spots and may not be able to support all organizations responding to a disaster.

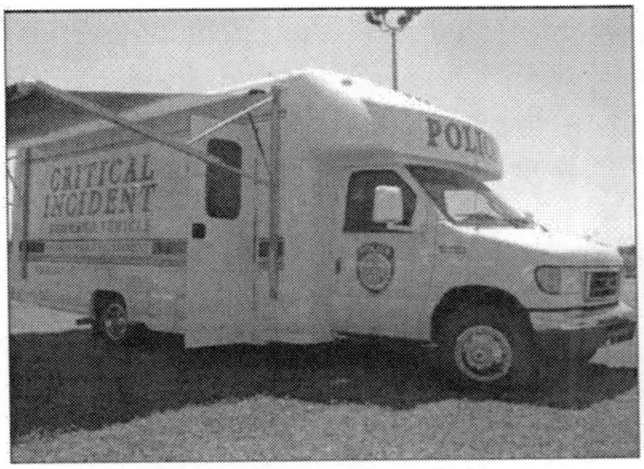

Jackson Police Department communications vehicle

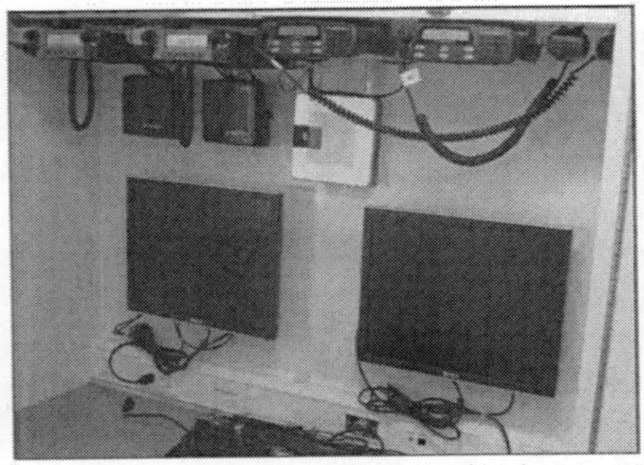

Interior of the Jackson Police communications vehicle, showing a work desk with two computer monitors and wireless Internet

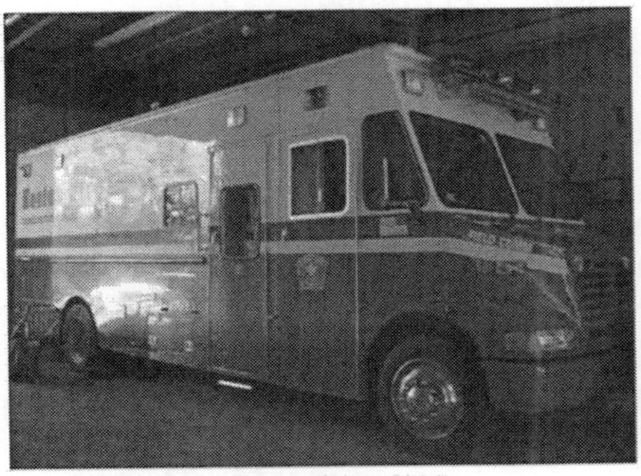

Boston Fire Department communications vehicle

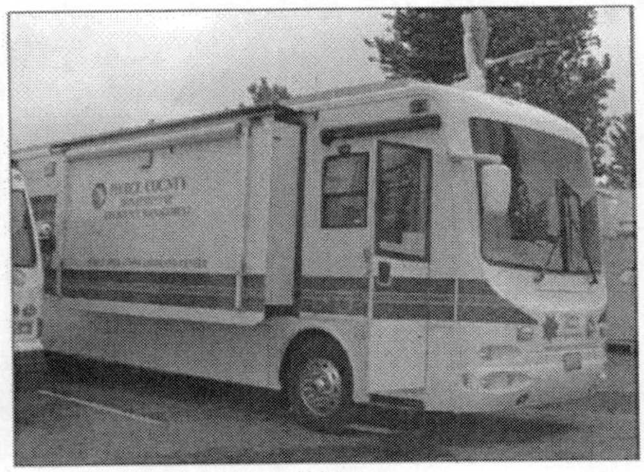

Pierce County communications vehicle

Source: GAO.

Figure 8. Jurisdictions' Emergency Response Vehicles.

Inability to Send Non-Voice Communications

Some current systems are not designed to send non-voice communications, such as photographs and video. First responders in several of our case study areas described additional capabilities that developing non-voice communications would provide. For example, photographs and video can quickly convey an emergency situation, saving time in response. Related capabilities, such as geospatial mapping, can accurately identify the location of first responders relative to a disaster. Hawaii first responders described an instance battling brush fires when air reconnaissance had to roughly convey its location using voice descriptions compared with a paper map. Some jurisdictions we interviewed are expanding, or planning to expand, their systems' capacity to transmit photographs and videos. For example, first responders in Sacramento, California, have planned to install a digital radio system in their emergency operations center, which would enable both voice and data transmissions.

INTEROPERABILITY VULNERABILITIES PERSIST

We have previously reported on vulnerabilities involving interoperability, which is the ability of first responders to communicate with whomever they need to (including personnel from a variety of agencies and jurisdictions), when they need to, and when they are authorized to do so. Facilitating interoperability has been a concern for many years, and we have cited a variety of obstacles to effective interoperable communications among first responders.[27] While we have reported on progress in improving interoperability among first responders, our case study work shows that technological and human factors continue to impair interoperability. Using the scenario of a hurricane hitting a coastal city, figure 9 depicts how interoperability may be threatened by jurisdictions using different technologies, fast-changing technologies, the critical nature of training, and jurisdictional tensions.

Jurisdictions Use Different and Fast-Changing Technology

First responders continue to use various, and at times, incompatible communications technology, making it difficult to communicate with neighboring jurisdictions or other first responders to carry out response. For example, some fire departments have hesitated to use digital radio systems due to safety concerns, which could create incompatibility with other responders' equipment, such as law enforcement (see app. III for more information on communications systems used by first responders).[28] According to first responders in Tennessee, Massachusetts, and Washington state, 800 MHz radio systems perform poorly in buildings. Difficult radio communications in high-rise buildings contributed to firefighter deaths during the September 11, 2001, terrorist attacks in New York City as some firefighters did not receive the transmission to evacuate the World Trade Center. In another example, Hawaii's geographic isolation has contributed to island jurisdictions independently designing their communications systems, resulting in disparate systems statewide. This can prove problematic for interoperability, particularly if a major disaster required responder assistance from neighboring islands.

Given the fast-changing nature of communications technology, upgrade needs and replacement cycles compound interoperability vulnerabilities. Officials at the National Public Safety Telecommunications Council reported that keeping up with technology is difficult for jurisdictions due to funding constraints. Yet some jurisdictions must upgrade when manufacturers eliminate technical support for older systems. Other legacy systems still in use are aging or obsolete. For example, some communications systems currently used by California's first responders have reached or exceeded their life expectancy, while other components need replacement. Not all jurisdictions, however, maintain the same upgrade schedule. For example, first responders in Pierce County, Washington, described coordinated efforts by them and other jurisdictions to help ensure that different technological upgrades and other system changes increase rather than reduce existing interoperability.

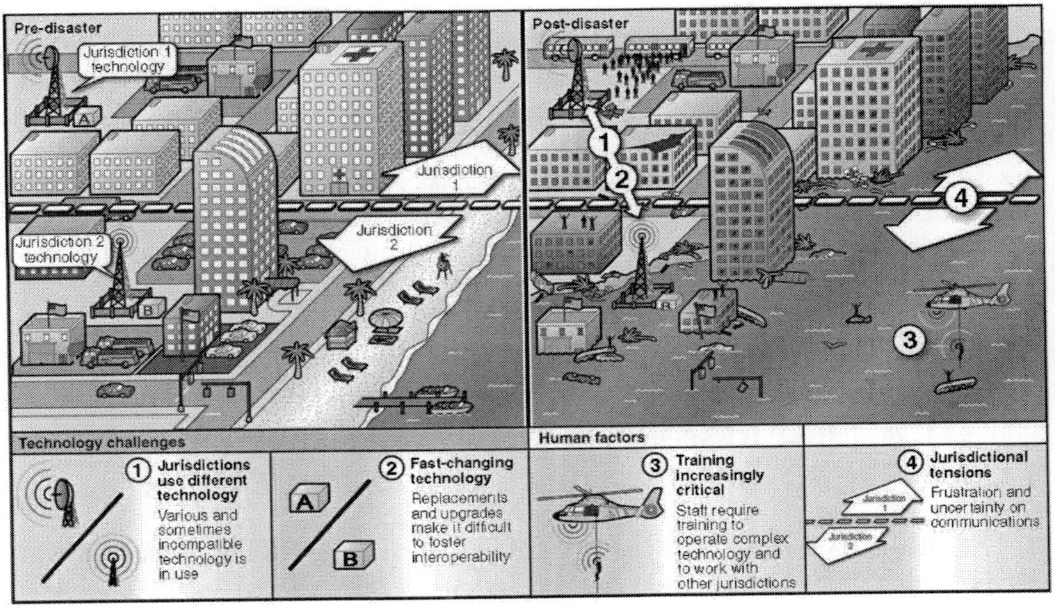

Source: GAO.

Figure 9. Vulnerabilities Involving Interoperability in a Hurricane Scenario.

To account for different and sometimes incompatible communications systems, some jurisdictions have used technologies to facilitate interoperability by "patching" together different systems into a common network. For example, first responders in Florida, Massachusetts, and Washington state described using equipment to create a local area network that can patch in different communications systems. This patched network can create local interoperability among different jurisdictions' communications systems. However, first responders noted instances where patching technology failed to establish interoperability.[29] Also, some patching equipment cannot provide blanket interoperability for an entire city or county and thus may be insufficient to meet communications needs in a catastrophic incident.

Human Factors

Several jurisdictions emphasized that training was increasingly critical to operate complicated equipment and coordinate with multiple jurisdictions to improve interoperability. The Massachusetts Executive Office of Public Safety and Security reported that achieving interoperability not only requires equipment, but staff must be regularly trained to work effectively with a number of jurisdictions. According to the 9-11 Commission Report,[30] the New York City Police and Fire Departments were not prepared to comprehensively coordinate with one another on the day of the September 11, 2001, terrorist attacks. This led to communications breakdowns where responding agencies lacked knowledge of what other agencies were doing. For example, firefighters did not receive information from police helicopters regarding damage to the World Trade Center. There were also jurisdictional tensions as some reports indicated that firefighters refused to evacuate when asked by police officers, contributing to deaths. We observed jurisdictional tension in several of our case studies, which could inhibit cooperation and achieving interoperability. First responders in Florida, Massachusetts, and Washington state noted frustration with neighboring jurisdictions and uncertainty over how jurisdictions would communicate in the event of a disaster. Working well with others and reducing tensions has taken on increasing importance as more jurisdictions, such as public works, are regarded as first responders and participate in emergency communications. Memphis first responders said that achieving interoperability requires not only compatible technology, but also jurisdictions building relationships among personnel.

A Wide Range of Federal Assistance Aimed at Helping First Responders Mitigate Emergency Communications Vulnerabilities

Federal agencies such as DHS and FCC have a wide range of assistance intended to help first responders mitigate emergency communications vulnerabilities. Available assistance includes federal agency guidance, grants, and technical support. We have identified several examples of key federal assistance used by first responders.[31]

New Strategic Guidance among Significant Federal Efforts to Enhance Emergency Communications

DHS and other federal agencies have recently developed strategic guidance and pursued significant efforts to enhance emergency communications. Efforts such as the National Emergency Communications Plan, the Emergency Communications Preparedness Center, and various stakeholder and advisory groups reflect an emphasis on developing a more strategic approach to federal government efforts to mitigate emergency communications vulnerabilities. Other recent efforts underway include FCC's new approach to establishing a 700 MHz Public/Private Partnership.

National Emergency Communications Plan

DHS's Office of Emergency Communications released the National Emergency Communications Plan in July 2008, providing a framework for emergency communications users across all levels of government.[32] The plan is the first strategic document focused exclusively on improving emergency communications nationwide, and outlines an overarching strategy to address emergency communications shortfalls for federal, state, and local first responders. The plan includes strategic emergency communications goals and objectives, and recommends numerous initiatives and milestones to guide emergency response providers and government officials in making measurable improvements to emergency communications (see figure 10). Congress required this plan—developed with federal, state, local, and private sector stakeholder involvement and multiple policy and planning documents—to be subject to periodic review and updates. An important foundation for the plan, the National Communications Capabilities Report—also released in July 2008—provides a framework for evaluating the emergency communications capabilities of federal, state, and local agencies and organizations, which, according to the report, vary.[33] Both of these reports build on the groundwork set by the 2004 DHS SAFECOM Interoperability Continuum, which recognizes the importance of a formal governance structure to ensure the success of interoperability planning, including improving the policies and procedures of major projects by enhancing stakeholder coordination and establishing guidelin and principles.

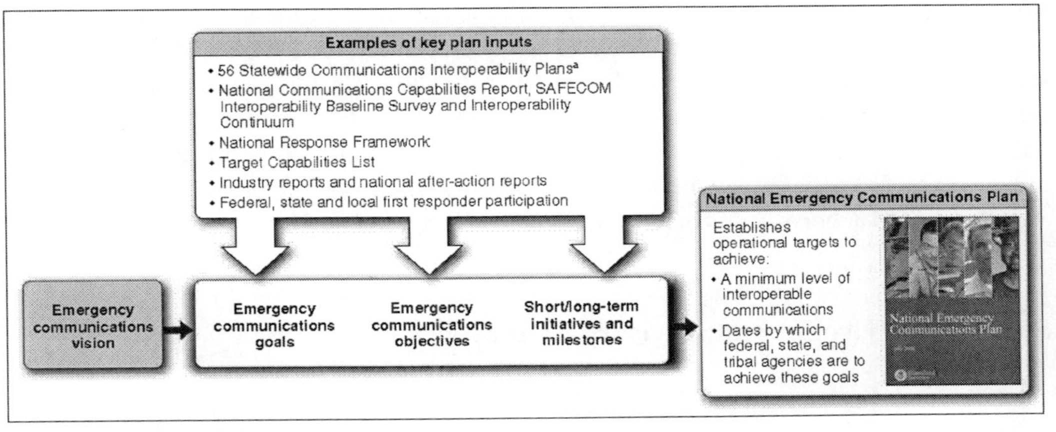

Source: GAO analysis of DHS information.

[a]The 56 Statewide Communications Interoperability Plans include the District of Columbia and 5 territories.

Figure 10. National Emergency Communications Plan Framework.

Emergency Communications Preparedness Center

The Post-Katrina Act requires federal agencies including DHS, FCC, DOJ, and the Department of Commerce to establish and jointly operate the Emergency Communications Preparedness Center.[34] Under the act, the center is intended to serve as the focal point and clearinghouse for intergovernmental emergency communications information sharing, and is required to submit to Congress an annual strategic assessment on federal coordination to advance emergency communications. The Emergency Communications Preparedness Center is to provide a governance and decision-making structure for strategic interagency coordination of emergency communications at the federal level. The center will not be officially established until a MOU has been finalized and approved by the signatory agencies.[35] DHS's Office of Emergency Communications chairs the Emergency Communications Preparedness Center working group. The working group drafted a charter, which defines the mission and roles of its members. Once approved, the charter will serve as the MOU governing the Emergency Communications Preparedness Center. As of June 2009, the agencies were working on completing the MOU.

700 MHz Public/Private Partnership

The FCC is pursuing a 700 MHz Public/Private Partnership to promote a nationwide interoperable broadband network for public safety that would increase the bandwidth capacity available for first responders in both day-to-day operations and during an emergency response. This has been a key FCC effort with regards to emergency communications and is a significant departure from prior FCC public safety spectrum allocations, which assigned spectrum licenses on a jurisdiction-by-jurisdiction basis. However, after the first attempt to auction the nationwide D Block license did not result in a winning bidder, FCC issued two further notices of proposed rulemakings, and a final order has not been adopted.[36] According to statements in FCC's Third Further Notice, a public/private partnership remains the best option to achieve nationwide build-out of an interoperable broadband network for public safety, given the current absence of federal appropriations for this purpose and the limited funding available to the public-safety sector. In April 2009, FCC officials reported that they were exploring ideas and options for future use of the spectrum. While the ultimate outcome of the 700 MHz Public/Private Partnership is currently unknown, the proceeding has involved significant FCC action over the course of several years (see table 2).

Stakeholder Groups and Advisory Committees

DHS and FCC have established stakeholder groups and advisory committees to help leverage existing knowledge and provide strategic recommendations to improve emergency communications. The purpose of these groups is to contribute expertise, recommendations, and lessons learned from recent disasters to help improve emergency communications. For example, the FCC's Independent Panel Reviewing the Impact of Hurricane Katrina on Communications Networks studied the effect of Hurricane Katrina on all sectors of the telecommunications and media industries, including public safety communications.[37] The panel then reviewed the sufficiency and effectiveness of the recovery effort and made recommendations to FCC regarding ways to improve disaster preparedness, network reliability, and communications. More detailed information on these emergency communications related groups and committees—including missions and activities—can be found in appendix IV.

Table 2. 700 MHz Public/Private Partnership Proceeding.

700 MHz Public/Private Partnership: Major Actions	
August 1997	Congress allocated 24 megahertz of spectrum in the Upper 700 MHz Band for public safety services.[a]
December 2006	FCC proposed a centralized and national approach to maximize public safety access to interoperable, broadband spectrum in the 700 MHz band, and to foster and promote the development and deployment of advanced applications (e.g., data and video), technologies, and systems.[b]
July 2007	FCC created a nationwide license in the D Block and required the winning commercial bidder to work with the Public Safety Broadband Licensee in a Public/Private Partnership—governed by FCC rules and a network sharing agreement—to construct and operate a nationwide network shared by commercial and public safety users.[c]
November 2007	The Public Safety Spectrum Trust was granted the license for the Public Safety Broadband Licensee.[d]
March 2008	FCC's Auction 73failed to attract a winning commercial bidder for the D Block of spectrum.[e]
September 2008	FCC proposed a modified set of rules and a revised auction plan inthe Third Further Notice, which includes a proposal to use the bidding process to determine whether the D Block spectrum would be licensed on a nationwide or regional basis.[f]
November 2008	The Third Further Notice public comment period closed, and FCC was continuing to review comments as of June 2009.

Source: GAO analysis of FCC information.

[a] See Balanced Budget Act of 1997, Pub. L. No. 105-33, 111 Stat. 251 § 3004 (1997) (adding § 337 of the Communications Act); Reallocation of Television Channels 60-69, the 745-806 MHz Band, Report and Order, 12 FCC Rcd 22953 (1998), recon. 13 FCC Rcd 21578 (1998).

[b] See Implementing a Nationwide, Broadband, Interoperable Public Safety Network in the 700 MHz Band, Development of Operational, Technical and Spectrum Requirements for Meeting Federal, State and Local Public Safety Communications Requirements Through the Year 2010, Ninth Notice of Proposed Rulemaking, 21 FCC Rcd 14837 (2006) (700 MHz Public Safety Ninth Notice).

[c] 700 MHz Second Report and Order, 22 FCC Rcd 15289 (2007).

[d] Implementing a Nationwide, Broadband, Interoperable Public Safety Network in the 700 MHz Band, Order, 22 FCC Rcd 20453 (2007).

[e] See Auction 73, 700 MHz Band, at http://wireless.fcc.gov/auctions/default.htm?job=auction _summary&id=73.

[f] 700 MHz Third Further Notice of Proposed Rulemaking, 23 FCC Rcd 14301 (2008). See, also 700 MHz Second Further Notice of Proposed Rulemaking, 23 FCC Rcd 8047 (2008).

Regional Emergency Communications Coordination Working Groups

As required by the Post-Katrina Act, and in an effort to develop a new regional governance structure, FEMA has recently established 10 Regional Emergency Communications Coordination Working Groups (see Figure 11), intended to work closely with federal, state and local officials to improve emergency communications.[38] Specifically, the working groups are to assess local emergency communications systems' ability to meet the goals of the National Emergency Communications Plan; facilitate disaster preparedness by promoting multi-jurisdictional and multi-agency emergency communications networks; and ensure activities are coordinated with regional emergency communications stakeholders. FEMA has proposed that the working groups be the single federal emergency communications coordination point for disaster response and interaction with state and local governments. Many of the established working groups are in early stages of development. For example, the Region X working group—covering Mount Rainier in Washington state— has held one stakeholder meeting. As of June 2009, all 10 of the FEMA Regions had established working groups. In addition, FEMA has hired 1 national and 10 regional positions to coordinate these working groups.

A VARIETY OF FEDERAL FUNDING AVAILABLE

Federal agencies have several grants available for states, territories, and local and tribal governments that are used for emergency communications. In 2008, interoperable emergency communications represented the largest investment category of DHS grants, including more than a dozen grant programs such as the Urban Areas Security Initiative,[39] the State Homeland Security Initiative,[40] the Interoperable Emergency Communications Grant Program,[41] and the Emergency Operations Center Grant Program.[42] FEMA, which is responsible for allocating and administering DHS grants, awarded over $3.85 billion in federal funding to improve interoperable emergency communications to state and local agencies from 2004 to 2007. FEMA manages the majority of federal grants for disaster preparedness and response; however, other federal agencies have contributed to this effort. The total amount of federal funds directed to emergency communications interoperability in the last 8 years is difficult to determine because after the September 11, 2001, terrorist attacks, multiple federal agencies offered funding to state and local governments in preparation for natural and man-made disasters. Interoperability was among several grant criteria for broad preparedness funds that could be used for a number of things, including interoperable emergency communications.

Historically, DOJ has also contributed to emergency communications efforts. Many first responders in our case study locations reported that they received funding to improve emergency communications from DOJ grant programs. For example, Boston, Honolulu, and Miami participated in DOJ's 25 Cities Project,[43] which funded initiatives to address communication networks between key state and local authorities in major metropolitan areas that were determined to be at a higher risk for terrorist attack. Boston, Memphis, and Sacramento also received funding from DOJ's Interoperable Communications Technology Program,[44] which funds local and regional voice and data interoperability projects. Between 2003 and 2006, the Community Oriented Policing Program invested over $250 million in 65 agencies to improve jurisdictions ability to talk across disciplines such as fire and police departments using radio communications networks. In 2007, DOJ awarded $5.7 million to the Sacramento Police Department to support technology projects facilitating voice and data information sharing.

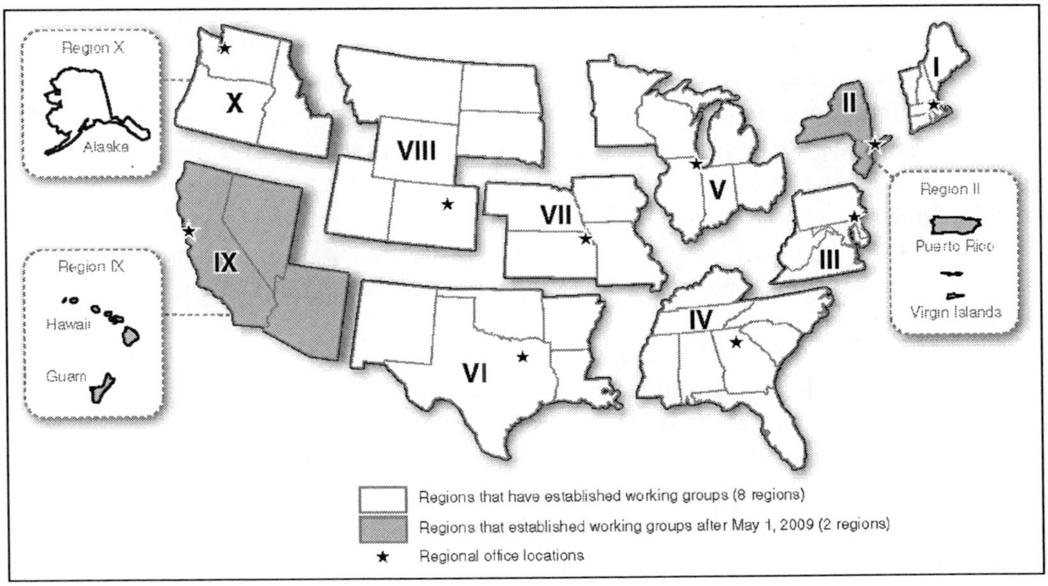

Source: FEMA (data); MapArt (map).

Figure 11. Status of Regional Emergency Communications Coordination Working Groups.

More recent federal funding has largely come from DHS and been focuse d on addressing specific gaps and identified needs, such as interoperable emergency communications. In 2007, all 56 states and territories received a portion of the approximately $1 billion, one-time Public Safety Interoperable Communications Grant Program funding to purchase hardware and update technology for interoperable communications systems. These funds were provided to assist public safety agencies in the planning and coordination associated with the acquisition of, deployment of, or the training for the use of interoperable communication equipment, software, or systems. The Public Safety Interoperable Communications Grant Program is an NTIA program. NTIA and DHS signed a joint collaboration agreement to have FEMA administer the grant program. This funding assists public safety agencies in improving communications through investments identified by each state or territory's Statewide Communications Interoperability Plan (SCIP), which FEMA required prior to release of grant funds.

The Post-Katrina Act required DHS to ensure consistency between grant guidelines and the goals and recommendations of the National Emergency Communications Plan. Requiring states to develop SCIPs was one step in an overall effort to align DHS-administered funding with the National Emergency Communications Plan. In developing the SCIPs, states involved local agencies and stakeholders to help identify communication and interoperability gaps to better address vulnerabilities (see Figure 12). The plans were developed using a methodology, which identified and developed working groups or governance councils to assure state-level accountability. For example, to ensure that local, regional, tribal, and state needs would be addressed and coordinated, California combined efforts of its existing Statewide Interoperability Executive Committee with other strategic planning groups. The Office of

Emergency Communications, FEMA, and NTIA jointly oversaw a peer review of the SCIPs and the investment justifications, using panels to review both documents, in order to ensure that Public Safety Interoperable Communications investment justifications addressed gaps that had been identified in the SCIPs. The Office of Emergency Communications used the recommendations from the peer review to approve the SCIPs in February 2008, and FEMA and NTIA used the information from the peer review to approve the investment justifications. All states where our case studies were located developed plans and received funding through the 2007 Public Safety In teroperable Communications grant[45] ranging from $8.1 million for Hawaii to $94 million for California. California is using some of these funds to pursue 16 statewide communications initiatives, including enhancing and implementing interoperability in the Sacramento area.

Beginning in 2008, FEMA and the Office of Emergency Communications worked together to develop the Interoperable Emergency Communications Grant Program. Whereas the Public Safety Interoperable Communication s Grant was a one-time investment, this new grant program is ongoing and is intended to help enable state, territorial, and local governments to implement their SCIPs. The program funds initiatives in governance, planning, equipment, training, and exercises that are consistent with the strategic initiatives and milestones outlined in the National Emergency Communications Plan. The Interoperable Emergency Communications Grant Program awarded $48.6 million in both fiscal years 2008 and 2009.

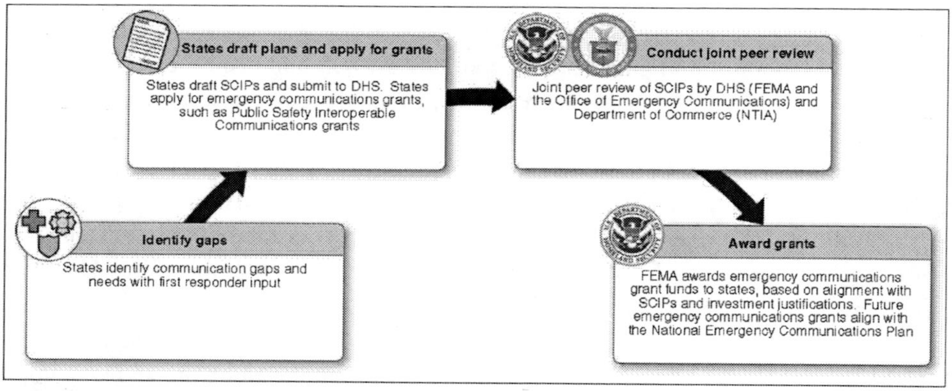

Source: GAO analysis of DHS information.

Figure 12. Public Safety Interoperable Communications Grants and Efforts to Align Targeted Investments for First Responders with the SCIP.

TECHNICAL SUPPORT AND FEDERAL ASSETS ARE INTENDED TO HELP MITIGATE EMERGENCY COMMUNICATIONS VULNERABILITIES

Several federal agencies with a role in disaster response offer technical support and initiatives in advance of an incident, or some can provide federal assets at the scene of a disaster to help mitigate emergency communications vulnerabilities. Federal agencies such as DHS and DOJ have developed technical support offerings intended to assist first responders in advanced planning and emergency preparedness. Similarly, in response to a real-time incident, DHS and FEMA can establish a physical presence at the disaster site, deploying personnel and assets to assist first responders. Technical support and planning provide assistance to address individual state and local jurisdictions' emergency communications needs. The following programs and efforts are examples of technical assistance and training available to assist first responders in improving continuity of communications, capacity, or interoperability, among other vulnerabilities.

- *Interoperable Communications Technical Assistance Program.* DHS's Interoperable Communications Technical Assistance Program provides support to first responders for planning and technical issues that need to be considered when first responders develop interoperable communications. The program supplies a site management team and support to each area requesting assistance, providing technical assistance and analysis tailored to meet site-specific requirements. All of our case study scenario states have received technical assistance and services through this program. For example, Hawaii received assistance on 17 work requests, including communications unit leader training, a tabletop exercise, [46] and engineering support. Many states also used this technical assistance to aid in the development of their SCIPs. As previously discussed, DHS and FEMA also provided feedback to assist states in completing these plans, as well as input to assure alignment with the National Emergency Communications Plan.
- *Catastrophic Disaster Response Planning Initiative.* In 2006, FEMA began a Catastrophic Disaster Response Planning Initiative combining planning and exercises to produce functional plans for areas at risk of a catastrophic disaster. In this ongoing effort, communications is one of several functional areas FEMA is addressing with state and local first responders. This involves planning for disaster scenarios—including a catastrophic earthquake in the New Madrid seismic zone and a hurricane in Florida—two of our case study locations. In the earthquake-planning scenario, for example, FEMA officials are focused on a bottom-up approach (i.e., beginning at the local level across all disciplines, then rolling up to the state level to identify gaps and craft the regional plan to mitigate those gaps) and completed 14 local workshops and 18 state-level workshops in 2008, which included approximately 3,800 stakeholders at all levels of government.

- *Government Emergency Telecommunications Service and Wireless Priority Service.* NC S's Government Emergency Telecommunications Service provides subscribers with access cards for priority service over wireline telephone networks in an emergency.[47] The FCC and NCS's Wireless Priority Service offers a similar service for cellular networks, and both of these services can be useful in mitigating capacity vulnerabilities when demand overwhelms communications systems immediately following an incident. State and local first responders in many of our case study locations participated in the Government Emergency Telecommunications Service program. NCS also manages the Telecommunications Service Priority Program, which provides national security and emergency preparedness users priority authorization of telecommunications services.

- *Integrated Wireless Network*. In 2001, DHS, DOJ, and the Department of the Treasury began a collaborative effort to develop the Integrated Wireless Network and provide secure, seamless, and interoperable wireless communications for federal agents and officers engaged in law enforcement, homeland defense, and disaster response. Initially conceived as a joint radio communications solution to improve communication among federal, state, and local law enforcement agencies, the Seattle/Blaine area in Washington state began a pilot network in 2004. While the pilot continues to provide service to multiple agencies, the departments have determined that this specific system design cannot be implemented on a nationwide scale. Consequently, the formal governance structure that was initially established among the three departments h as been disbanded, and the contract for developing a new design is not currently being used jointly by the departments for this purpose.

- *Project 25*. The Association of Public Safety Communications Officials' Project 25 is a long-standing effort to select common system standards for digital public safety radio communications.[48] These standards are intended to allow radios to be interoperable regardless of manufacturer. We have previously reported that implementation of systems based on incomplete Project 25 standards has been problematic.[49] With no process in place to confirm that equipment advertised as compliant actually met the standards, Congress called for the creation of the Project 25 Compliance Assessment Program.[50] This voluntary program establishes a process for equipment suppliers to submit their equipment to certain testing labs to receive a certification of Project 25 compliance.[51]

In response to a disaster, federal assets are also available on the ground to help mitigate one or more emergency communications vulnerabilities, including continuity of communications, capacity, and interoperability. Some emergency response personnel and equipment may be deployed to the scene, such as DHS and FEMA officials, while other federal agencies may have personnel at the scene based on the nature and/or location of the incident. For example, at Mount Rainier National Park in Washington state, National Park Service personnel physically located on site would be directly involved in any response effort taking place within the park. Some federal agencies have assets available that can be deployed during or immediately following an incident and can help mitigate continuity of communications vulnerabilities. For example, FEMA maintains 6 deployable Mobile Emergency Response Support detachments across the country. These detachments provide personnel, vehicles, and technology on the scene and can help other federal agencies, state, or local first responders establish communications. Mobile Emergency Response Support detachments can be activated at the request of state authorities to provide communications on the scene when existing state and local communications infrastructure has been damaged or destroyed. For example, some of the vehicles in the detachment have the communications equipment necessary to facilitate full voice, data, and video multi-agency interoperability and can operate as a stand-alone communications center. The Maynard, Massachusetts, detachment was deployed 41 times in 2007, 34 times in 2008. Based on its proximity t o one of our case study locations, the Bothell, Washington, detachment could be an effective tool for restoring communications after a catastrophic disaster at Mount Rainier, as equipment and personnel could arrive on scene within 12 hours after an incident (see figure 13).

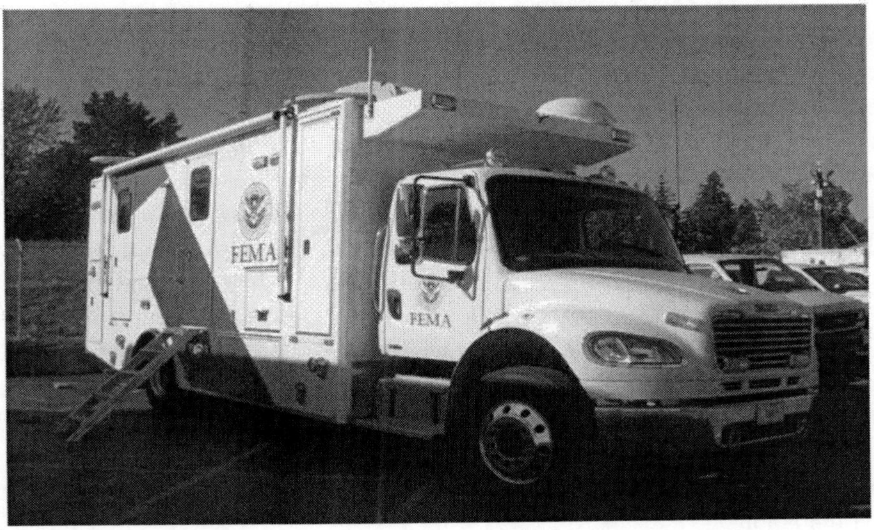

Source: GAO.

Figure 13. FEMA Mobile Emergency Response Support Vehicle.

LIMITED COLLABORATION AND MONITORING JEOPARDIZE SIGNIFICANT FEDERAL EFFORTS AND IMPEDE PROGRESS

Limited collaboration and monitoring impedes the progress of some significant efforts being undertaken by federal agencies to strategically enhance emergency communications. Our past work has shown, and the National Emergency Communications Plan articulates, that collaboration and monitoring are important elements to advancing emergency communications.[52] We found that federal agencies have demonstrated limited application of collaboration best practices, as well as lack mechanisms for fully monitoring efforts.

Collaboration Key to Advancing Emergency Communications

Defining a common goal and mutually reinforcing strategies are collaboration best practices that can help federal agencies deal with issues that are national in scope and cross agency jurisdictions.[53] In particular, establishing a governance structure that includes defined leadership, roles, and responsibilities can be an effective step for establishing goals and aligning strategies so that they are mutually reinforcing.[54] Addressing goals by leveraging resources is another collaboration best practice that can be employed across agencies to maximize resources. The National Emergency Communications Plan acknowledges the importance of collaboration, including at the federal level, for enhancing emergency communications. Among the plan's seven objectives is that federal emergency communications programs and initiatives be collaborative and aligned to achieve national goals. Additionally, the plan speaks to the importance of federal programs and initiatives related to emergency communications being coordinated so as to minimize duplication, maximize federal investments, and ensure interoperability.

Strong collaboration is especially important since DHS has limited authority to compel other federal agencies to participate or align their emergency communications activities despite DHS's leadership role in compiling and overseeing the National Emergency Communications Plan. DHS officials noted that the agency cannot unilaterally achieve the strategic goals, initiatives, and milestones of the National Emergency Communications Plan and will rely on the voluntary commitment of federal, state, local, and tribal government officials and the private sector.[55] DHS's tools to encourage the participation of stakeholders include the technical assistance that DHS's Office of Emergency Communications provides to state, regional, local, and tribal government officials and the development of grant policies that align with the National Emergency Communications Plan. The agency has fewer instruments to encourage federal agencies' participation. For example, the National Communications Capabilities Report notes that federal agencies are not eligible to receive grants, and therefore, grant guidance is an ineffective means for encouraging and coordinating federal participation.

Establishing Common Goals and Mutually Reinforcing Strategies Can Enhance Some Significant Emergency Communications Efforts

In our past and current work, we found that federal agencies have demonstrated limited application of certain collaboration best practices with respect to some significant federal efforts.[56] Limited collaboration has contributed to the failure of past federal emergency communications efforts and puts ongoing efforts at risk.

Integrated Wireless Network

In December 2008, we reported that federal efforts to provide a joint agency solution for federal emergency communications through the Integrated Wireless Network had not been successful, because federal agencies did not effectively identify a common goal and design mutually reinforcing strategies.[57] We found that success depended on a means to overcome differences in agency missions and cultures and a joint strategy to align activities and resources to achieve a joint solution. More specifically, DHS, DOJ, and the Department of the Treasury did not establish an effective governance structure. In abandoning collaboration as a joint solution, DHS, DOJ, and the Department of the Treasury risk duplication of effort and inefficient use of resources as they continue to invest significant resources on independent solutions. Further, the efforts being pursued by these three agencies will not ensure that vulnerabilities involving interoperability are addressed. We have previously reported that interoperability with federal first responders remains an important element in achieving nationwide interoperability, and until a federal coordinating entity makes a concerted effort to promote federal interoperability with other governmental entities, overall progress in improving communications interoperability will remain limited.[58]

Emergency Communications Preparedness Center

Delays in establishing the Emergency Communications Preparedness Center undermine implementation of the National Emergency Communications Plan's wide array of strategic goals, initiatives, and milestones scheduled to occur concurrently—most of which within the first year from the plan's July 2008 issuance—and depend on cooperation from multiple agencies to achieve. The National Emergency Communications Plan describes that milestones detail the timelines and outcomes of each of the 29 initiatives to serve as the "key checkpoints" to monitor the plan's implementation. To that end, the plan includes 11 January 2009 milestones, 2 April 2009 milestones, and 41 July 2009 milestones. These milestones scheduled to occur within the first year of the plan's issuance comprise more than half of the plan's total 91 milestones. As previously discussed, the Post-Katrina Act envisions the center will serve as a focal point for federal interagency coordination, and the National Emergency Communications Plan articulates that the center will help ensure that the strategic goals, initiatives, and milestones of the plan are agreed upon and that federal agencies work collaboratively to pursue mutually reinforcing strategies. The Office of Emergency Communications within DHS chairs the Emergency Communications Preparedness Center working group, which has drafted a MOU currently under review. As we have previously reported, an important element of establishing effective collaborative relationships is to reach formal agreements with each partner organization on a clear purpose, expected outputs, and realistic performance measures, which the center's MOU could supply.[59] In September 2008, DHS officials reported to Congress that the center's MOU was going to be completed by December 2008.[60] To date, the MOU has not yet been finalized. DHS officials reported to us that they are taking steps to establish the Emergency Communications Preparedness Center, but noted that the complexity of obtaining agreement from multiple agencies was a challenge. In the absence of the Emergency Communications Preparedness Center's finalized MOU, according to DHS officials, staff-level working groups are working to help implement the plan. With the final nature of the center as yet undetermined, however, FEMA officials expressed that many federal agencies and components are still functioning in an independent manner, which can be confusing to state and local first responders. Moreover, these officials' understanding of the center was that it would largely serve as a Web site clearinghouse for information. Other DHS officials reported that the center would serve a range of functions for emergency communications.

DHS Efforts and FCC's 700 MHz Public/Private Partnership

Collaboration between DHS and FCC on the 700 MHz D Block spectrum has been limited. Spectrum is a valuable resource for public safety wireless communications, and the 700 MHz spectrum that is becoming available as the result of the digital television transition represents a significant increase for public safety purposes. Limited collaboration jeopardizes the viability and usefulness of this spectrum for public safety and its relation to other federal efforts. By employing collaboration best practices, DHS and FCC could enhance what is ultimately done with the 700 MHz broadband spectrum and help accomplish the goals of the National Emergency Communications Plan.[61]

DHS and FCC officials do not have a common vision for the 700 MHz spectrum and have expressed varying views on the relationship between the 700 MHz Public/Private Partnership—FCC's current proposal for how the 700 MHz should be allocated and assigned—and the National Emergency Communications Plan.[62] According to DHS officials, FCC's proceeding to establish the 700 MHz Public/Private Partnership directly supports the goals of the National Emergency Communications Plan and FCC officials said that they believed their effort is compatible with the National Emergency Communications Plan. However, FCC officials also reported that they believed the National Emergency Communications Plan applied to only existing emergency communications systems and, therefore, was not directly relevant to allocating and assigning the spectrum, which they believed to be the main purpose of the 700 MHz Public/Private Partnership proceeding.[63] FCC officials described the plan and the 700 MHz Public/Private Partnership proceeding as two separate and parallel efforts.[64] These officials reported that, accordingly, it was reasonable that the proceeding's notices and other documents did not reflect or reference the National Emergency Communications Plan.

DHS's Office of Emergency Communications Director reported that it was too early for the office to have any significant role in developing the 700 MHz Public/Private Partnership, because the auction had not been completed.[65] DHS did not submit formal comments in this proceeding.[66] FCC officials stated that the lack of formal comments and critique from DHS suggested that the agency had no objections to the proposed rulemaking, and that it reflected the separate nature between the proceeding and DHS's efforts. FCC officials could not recollect nor provide us with a record of substantive conversations with DHS officials on this proceeding. However, the Director of DHS's Command, Control and Interoperability Division reported conveying to FCC officials several challenges regarding its 700 MHz Public/Private Partnership network before FCC's most recently issued document describing the proposed network, the Third Further Notice, was released. We recognize that FCC considers input from stakeholders, but also acknowledge DHS's important role in federal efforts regarding emergency communications. We compared the challenges this official expressed to any treatment in the Third Further Notice. Our analysis shows that on these issues that DHS and FCC do not share a common view in support of a 700 MHz Public/Private Partnership to build a nationwide, interoperable broadband network.[67] (See table 3.) We recognize that FCC's efforts are ongoing and that no final decision relative to the future of the 700 MHz public safety broadband spectrum has yet been made.

According to FCC officials, it would be the responsibility of the Public Safety Broadband Licensee to ensure that the network aligns with and furthers the goals of the National Emergency Communications Plan when it negotiates the details of the network sharing agreement with the winning commercial bidder. According to FCC officials, the Public Safety Broadband Licensee must ultimately meet the provisions outlined in the final rules adopted by FCC. It is unclear, however, how the Public Safety Broadband Licensee would do so or if it would have the authority to require such an alignment between the proposed network and DHS efforts. FCC did not suggest or specify in the Third Further Notice that the Public Safety Broadband Licensee consider the strategic goals and milestones of the National Emergency Communications Plan. Also, FCC did not in the Third Further Notice solicit comments on the relationship between the 700 MHz Public/Private Partnership and other DHS efforts. For example, FCC outlined no role for Regional Emergency Communications Coordination Working Groups, despite potential overlap with the 700 MHz Public/Private Partnership.[68] Specifically, the Public Safety Broadband Licensee is responsible for representing public safety interest in negotiating the network sharing agreement. In comparison, as legislatively defined and described within the National Emergency Communications Plan, the working groups are comprised of public safety officials who will assess emergency communications capabilities within their respective regions, facilitate disaster preparedness through the promotion of multijurisdictional and multiagency emergency communications networks, and serve as a primary link in coordinating multistate operable and interoperable emergency response initiatives and plans among federal, state, local, and tribal agencies. These working groups could provide a valuable means for representing public safety interests regionally and coordinating the use of a nationwide, interoperable, public safety broadband network envisioned by FCC.

Table 3. DHS Command, Control and Interoperability Division -identified Challenges to FCC's 700 MHz Public/Private Partnership.

Challenges identified by DHS Command, Control and Interoperability Division	FCC Third Further Notice
Pursuing a business model that requires subscriptions would be problematic for jurisdictions that lack funds.	FCC proposed a standard charge of $7.50 per month per user (meaning per public safety officer/individual) for gateway-based access to the shared network(s).
Local jurisdictions hesitate to spend more money to buy additional equipment, since they have already spent billions in newer communications equipment and infrastructure that must be maintained for mission-critical voice communications.	FCC recognized that "multi-band radios couldbe developed, although at some cost…that are capable of operating on both the shared wireless broadband network and other public safety frequency bands." Additionally, FCC "tentatively concluded to require the Upper 700 MHz D Block licensee to offer gateway-based access . . ." but proposed that "public safety users themselves bear the costs of thebridges and gateways, including installation and maintenance costs…"
Distributing any associated equipment in a disaster situation would pose a logistical challenge, if not be impossible.	FCC did not address this issue of how associated equipment needed in a disaster area would be dispersed, but sought comment on whether "it should require use oravailability of multi-band radios that could be available to public safety first responders that may need to come into these areas in times of emergency…"
Training and maintaining skills in using the network would be challenging, as first responders would need to use and exercise with this network often to maintain their familiarity with it.	FCC did not address or seek comment on if or how the proposed system would be made available for training and exercises.a

Source: GAO analysis of DHS and FCC information.

[a] According to FCC officials, mandating training requirements for state and local entities is outside FCC's jurisdiction. However, FCC officials also reported that it was possible for FCC to set conditions for the licensees operating the 700 MHz Public/Private Partnership to address such issues through its rulemaking process.

Based on further analysis of the Third Further Notice and interviews with FCC and DHS officials, we found potential opportunities to align the proposed 700 MHz Public/Private Partnership network with the National Emergency Communications Plan and other DHS efforts to reinforce one another.[69]

- Given the National Emergency Communications Plan's focus on Urban Area Security Initiative regions and the national planning scenarios and their continued use by DHS and FEMA, FCC could align the performance benchmarks to prioritize Urban Areas Security Initiative regions or reference in their definition of "emergency," DHS and FEMA's national planning scenarios.[70] FCC's 700 MHz Public/Private Partnership currently has no relationship to these efforts.
- FCC's Third Further Notice does not propose a specific role for state government in coordinating their public safety providers' participation in the 700 MHz Public/Private Partnership or the network, but state governments have played a key role in emergency communications by coordinating and completing SCIPs, which have been reviewed by DHS and were required to be eligible for federal funding.[71] FCC could consider ways to involve state government or integrate the use of SCIPs— particularly if regional licenses are awarded, because the majority of Public Safety Regions that FCC has proposed as the geographic regions are delineated along state lines.
- To help facilitate interoperability among federal agencies, the National Emergency Communications Plan outlines an initiative to implement the Advanced Encryption Standard for federal responders. FCC proposed no requirement that federal government agencies be provided access to the network and outlined no technical specification for the Advanced Encryption Standard.[72] Given that federal agencies play a key role during catastrophic events, if the 700 MHz network does not incorporate this standard,[73] it could pose a challenge for federal responders working with state and local responders on this network. The encryption of the network once built may not adhere to the Advanced Encryption Standard. Thus, if federal responders needed to share sensitive or classified information with one another or with local responders, this network might not be an option for them, since it may not meet their encryption standard. FCC could consider including this standard in the specifications for the network or ask the licensees to examine the consequences for not adhering to this standard and any potential remedies.
- FCC proposed that the network be based on a modern Internet Protocol platform and that interconnectivity through gateways and bridges be allowed. FCC officials told us that one of the key benefits of the network would be nationwide interoperability facilitated through this common technological architecture. According to DHS's Interoperability Continuum, though, interoperability is not the result of solely technological solutions. In order to achieve interoperability, other elements such as standard operating procedures, usage, governance, and training and exercises, must be addressed. FCC has not addressed these other elements. According to FCC officials, these facets of emergency communications fall under the purview of DHS and FCC has not contemplated requiring the licensee(s) to take any actions that would associate or connect the proposed network with DHS efforts. FCC officials also reported that historically, FCC has not taken such action within its rulemaking

process. FCC could require that the licensees adopt use of DHS's Interoperability Continuum as a framework for negotiating the terms of the network sharing agreement (see figure 14).

The lack of commonly defined goals for the 700 MHz spectrum and mutually reinforcing strategies with DHS efforts threatens the usefulness and viability of the network for public safety. There is some support for a 700 MHz Public/Private Partnership to build a nationwide, interoperable broadband network from entities such as the Association of Public-Safety Communications Officials, but officials from major metropolitan areas, such as New York City, reported concerns with how the network will be governed, as proposed in the Third Further Notice. Specifically, the Deputy Chief from the New York City Police Department expressed concern to FCC Commissioners in a July 2008 public hearing regarding a commercial vendor managing the network and said that his agency would likely not participate, because the network would not meet all their mission requirements.[74] First responders we met with in Boston, Seattle, and Sacramento expressed similar concerns about the proposed network. For example, Boston Police Department officials told us that based on their experience with commercial telecommunications providers and the proposed fees for using the network, they preferred to directly manage and control the spectrum's use in their jurisdiction. Should these and other jurisdictions not participate in the 700 MHz Public/Private Partnership's network, first responders in those areas would be left without access to a potentially vital resource during a catastrophic event, which is contrary to FCC's stated goals for the network.

Opportunities for alignment	**FCC 700 MHz Third Further Notice** 	**National Emergency Communications Plan and other DHS efforts** 
Performance benchmarks and strategic goals	FCC requires the D Block licensee(s) to provide signal coverage and offer service to at least: • 40 percent of the population in each Public Safety Region by the end of the fourth year, • 75 percent by the end of the tenth year, and • D Block licensee(s) will be required to meet other benchmarks after 15 years.	The National Emergency Communications Plan has strategic goals that include: • 90 percent of all high-risk urban areas designated within the Urban Area Security Initiative are able to demonstrate response-level emergency communications within one hour for routine events involving multiple jurisdictions by 2010, and • 75 percent of all jurisdictions are able to demonstrate response-level emergency communications within 3 hours, in the event of a significant incident as outlined in national planning scenarios by 2013.
Regional organization	58 Regional licenses designated by Public Safety Regions and one nationwide licensee (Public Safety Broadband Licensee) that would represent public safety in negotiating the network sharing agreement among other responsibilities.	10 DHS / FEMA Regional Emergency Communications Coordination Working Groups.
Role of state government	FCC proposed no specific role for state governments in coordinating their public safety providers' participation in the interoperable shared broadband network.	States have been instrumental in developing SCIPs, which are reviewed by DHS and NTIA and required by Congress to be eligible for federal emergency communications funding.
Federal agency access and encryption standard	FCC leaves access for federal agencies to the discretion of the Public Safety Broadband Licensee and outlines no technical specifications for the Advanced Encryption Standard.	Federal agencies will implement the Advanced Encryption Standard.
Interoperability	FCC interoperability through an Internet Protocol-based architecture.	DHS's SAFECOM Interoperability Continuum describes 5 elements for achieving interoperability: Governance, Standard Operating Procedures, Technology, Training and Exercises, and Usage.

Source: GAO analysis of FCC and DHS data.

Figure 14. Analysis of FCC's Third Further Notice and DHS Efforts.

DHS Could Leverage Emergency Communications Planning Expertise for Federal Agencies

As the federal government's lead agency on emergency communications, DHS has provided technical assistance and guidance on how to develop emergency communications plans, but these resources have primarily focused on state and local jurisdictions and less so on federal agencies. As previously discussed, DHS has provided extensive assistance to state and local jurisdictions in developing emergency communications plans, which have been valuable to state and local first responders in understanding their communications capabilities and limitations and working toward further enhancements. DHS has also issued guidance on emergency communications planning directed at all levels of government. For example, the National Preparedness Guidelines/Targeted Capabilities List, issued in September 2007, included 13 critical tasks for how to "Develop and Maintain Plans, Procedures, Programs, and Systems." DHS guidance outlines the importance and some key elements of emergency planning for preparedness and response, such as consideration of the systems that will be used, personnel (those who can use these systems), and other relevant considerations.

Not all federal agencies have developed communications plans or conducted communication infrastructure threat and vulnerability assessments, making their preparedness to assist state and local first responders uncertain.[75] According to the National Emergency Communications Plan, few agencies conduct communications infrastructure threat and vulnerability assessments as part of emergency communications planning on critical communications assets. As previously discussed, the National Communications Capabilities Report notes that some federal agencies currently have no formal plans in place to identify communications vulnerabilities or map a way forward to mitigate such vulnerabilities.

DHS has begun efforts that could assist other federal agencies, but it is unclear whether these will result in formal emergency communications plans. The National Emergency Communications Plan observes that some agencies do not view emergency communications planning as a priority and includes a milestone that by July 2009 DHS will develop "a standardized framework for identifying and assessing emergency communications capabilities nationwide." The plan also includes a milestone for DHS providing "best practices and methodologies that promote the incorporation of vulnerability assessments as part of emergency communications planning." Each federal agency and department should also assess its existing communications capabilities and compare them to the capabilities needed to complete each agency's missions. According to DHS officials, they are taking steps to help meet these milestones and to outreach and assist other federal agencies. However, it is too early to tell what direction these new efforts will take.

A formal emergency communications plan can help federal agencies respond in a catastrophic disaster by enhancing agencies' understanding of their emergency communications capabilities. Without such planning and understanding, federal agencies' fundamental readiness and response declines, including their ability to support state and local first responders in disaster. For example, FEMA Region IX officials reported that planning activities with various California jurisdictions have helped increase understanding on how FEMA will work and support California first responders, as well as achieving a common understanding of communications operations and assets. We have previously reported that limited emergency communications planning has reduced the federal government's readiness to support state and local first responders and contributes to poor response. The Hurricane Katrina Lessons Learned Report highlighted that communications problems due to limited planning had a debilitating effect on response efforts in the Gulf Coast region.[76] Specifically, many available communications assets were not utilized fully because there was no national, statewide, or regional communications plan that incorporated them. According to a senior DHS official, agencies may find themselves at the center of response in certain disasters or other events, at which time communications weaknesses are revealed. Officials from federal agencies in our case studies, such as National Park Service officials at Mount Rainier and FBI officials, also reported the importance of emergency communications planning in preparedness and response to a catastrophic event. For example, FBI officials and local first responders in our Boston terrorism case study would be required to work closely together. According to a senior FBI official, a number of issues could interfere with the ability of FBI agents to carry out their duties, such as a lack of interoperability with local radio systems, inadequate encryption, and insufficient coverage.

Limited Monitoring May Impede Progress in Emergency Communications

We found that DHS and FCC had only limited processes in place to monitor and evaluate recommendations from stakeholder groups. We have previously reported that monitoring and evaluating efforts are crucial elements to achieving agency goals. Following up on stakeholder group recommendations could help key decision makers within the agencies to obtain feedback for improving both policy and operational effectiveness.[77] Instituting some of the recommendations these groups have made may improve emergency communications. For example, the National Coordination Committee, an advisory group set up by FCC, recommended that FCC require standard channel nomenclature for all interoperability channels in 2003.[78] During disaster response, it is crucial that all responding public safety agencies are able to tune their radios to the frequency or frequencies that the incident commander directs. However, there is little uniformity in the naming of radio channels—some responders designate their channels by colors, others by numbers. Standardized channel nomenclature could enhance interoperability, since responders across different jurisdictions and disciplines would use identical terminology for identifying radio frequencies, thereby minimizing confusion and delay. Standard channel nomenclature could prove particularly useful in catastrophic disaster response, because of the large numbers of responders from different jurisdictions and disciplines. According to the National Public Safety Telecommunications Council, FCC has not adopted this recommendation made in 2003. FCC officials reported that FCC made reference to this stakeholder group recommendation in a 2006 proceeding. FCC officials indicated that the recommendation was raised again in the ongoing 700 MHz Public/Private Partnership proceeding and that the recommendation would be addressed therein.

Without monitoring and evaluation, it is unclear how DHS and FCC have incorporated or kept pace with the work of their stakeholder groups. As previously discussed, the constantly evolving nature of emergency communications can create opportunities and challenges, some of which advisory groups have addressed. According to the National Public Safety Telecommunications Council, significant progress has been made in implementing recommendations that contribute to improved emergency communications, but meeting the demand for public safety communications is a dynamic process requiring ever-additional work. Our analysis revealed that stakeholder groups assembled by DHS and FCC have made some recommendations repeatedly that could address identified vulnerabilities (see figure 15).[79]

Both agencies have various ways of examining some stakeholder group recommendations, but neither includes a mechanism to systematically monitor and evaluate all recommendations from stakeholder groups, or the agencies' response. As previously discussed, DHS's approach has been practitioner-driven. According to DHS officials, the agency tracks recommendations and input provided by its stakeholder groups in varying ways (see table 4).

Vulnerability addressed	Recommendation	Year of recommendation				
		2004	2005	2006	2007	2008
Continuity of communications	Multiple access methods and alternative communication technologies, so that emergency communications are not disrupted	FCC		FCC DHS	DHS FCC	
	Emergency Responder Classification/ Credentialing for Telecommunications Provider			DHS FCC	FCC	
	Pre-positioned or deployable communication assets, such as mobile radios and mobile cell towers for both public safety and commercial communications providers; related planning, training, and exercises.		FCC	FCC	FCC	
Capacity	Communication platforms capable of integrating different data types (voice, photos, etc.)	FCC			DHS	FCC
	Increased bandwidth to enable transmission of video and large format files such as blueprints and video files	FCC			DHS	FCC
Interoperability	Internet Protocol for an interoperable network of networks such as LMR, cellular, and wireline, which can also handle multiple services and applications	FCC	FCC		DHS	FCC
	Migration path from legacy systems to the Internet Protocol internetwork	FCC				FCC
	Develop Interoperability Rules to handle issues such as governance and prioritization	FCC			DHS	

Sources: GAO analysis of DHS and FCC stakeholder recommendations made in the following reports: Network Reliability and Interoperability Council VII reports #1 and #3; and Focus Group 1 B National Security Telecommunications Advisory Committee Issue Review; FCC Hurricane Panel Order--Report Recommendations of the Independent Panel Reviewing the Impact of Hurricane Katrina on Communications Networks; National Security Telecommunications Advisory Committee Report to the President on Emergency Communications and Interoperability; and Joint Advisory Commitee on Communications Capabilities of Emergency Medical and Public Health Care Facilities Report to Congress.

Figure 15. Analysis of Advisory Group Recommendations 2004-2008.

Not all these activities, however, result in the agency monitoring recommendations or evaluating its response. DHS's activities to monitor National Security Telecommunications Advisory Committee recommendations appear the most robust, as these mechanisms can account for which recommendations were acted upon and, if not, why. In contrast, though DHS tracking activities for its other stakeholder groups document recommendations and other information produced, DHS does not collect or record the agency's response.

Table 4. DHS Stakeholder Groups and Tracking Activities.

DHS stakeholder group	Tracking activity
SAFECOM Executive Committee and Emergency Response Council	DHS tracks the recommendations and input provided by the SAFECOM Executive Committee (EC) and Emergency Response Council (ERC): • Formally through meeting reports • Informally through detailed minutes taken during biannual ERC meetings, monthly and quarterly EC conference calls, and face-to-face meetings, as well as during ad hoc working group meetings. • For biannual ERC meetings, DHS develops a formal meeting report, which documents the key content, input, and recommendations from the working sessions.
National Security Telecommunications Advisory Committee	The National Communications System (NCS), within DHS, tracks recommendations made by the National Security Telecommunications Advisory Committee (NSTAC). Once a recommendation is approved by the NSTAC, which is typically on a quarterly basis, DHS/NCS convenes a team of NCS managers and subject matter experts to determine what priority area the recommendations fall under and provides a quarterly status report to the NSTAC Chairperson on whether the recommendation will be: • Taken for NCS action; or • Closed because: • There are insufficient resources; • The recommendation is overcome by events; • The recommendation is being addressed by another organization; or • The NCS has completed its activities.
Federal Partnership for Interoperable Communications	Federal Partnership for Interoperable Communications (FPIC) standing committees (Interoperability, Security, Spectrum, and Standards) conducts the following activities: • Identify potential recommendations; • FPIC committee members draft recommendations and submit to the FPIC general membership for review and formal acceptance, if needed; and • Any member agency or advisory member that disagrees with a decision or vote of the FPIC may submit a Minority Report.

Source: GAO analysis of DHS information on a selection of stakeholder groups and tracking activities.

FCC has not systematically monitored or evaluated recommendations of its advisory committees[80] and the agency's response, limiting the relevance of these groups' work. In December 2004, we reported that FCC did not have a process for tracking all its advisory committee recommendations. At the time, the deputy committee management officer told us that as a result of our review, FCC planned on improving the accountability of the advisory committee process by requiring committee recommendations be tracked. [81] To date, FCC has not established and instituted any such tracking mechanism. FCC officials reported that many of the recommendations from its advisory groups are not directed at FCC, and consequently, tracking or monitoring is less necessary. We note, in January 2009, the FCC Acting Chairman said that the agency could more fully take advantage of the work of advisory committees to increase agency transparency. FCC has tools at its disposal that it could use for monitoring and evaluating recommendations. For example, for its Independent Panel Reviewing the Impact of Hurricane Katrina on Communications Networks, FCC issued a Notice of Proposed Rulemaking (notice) in June 2006 inviting comment on what actions the Commission should take to address the Katrina Panel's recommendations.[82] FCC received over 100 comments and reply comments in response to the notice. On June 8, 2007, FCC released an order directing its Public Safety and Homeland Security Bureau to implement and track several of the recommendations that included FCC's rationale and conclusions behind selecting these particular recommendations (FCC Hurricane Panel Order).[83] The Public Safety and Homeland Security Bureau fulfilled the order by taking actions to implement the recommendations and reported to the commissioners after 3 and 9 months, as directed, on its actions.

Both DHS and FCC are forming new stakeholder groups, but no mechanisms to monitor or evaluate these groups' work, or the agencies' response, are currently in place. Without such mechanisms, it will remain unclear how and to what extent federal agencies have considered, or incorporated, the information provided by these groups. For example, as previously discussed, FEMA has formed Regional Emergency Communications Coordination Working Groups across the country. However, while there are defined roles and responsibilities for these working groups in both the legislation and the National Emergency Communications Plan, there are no legislative requirements for DHS, FEMA, or any other agencies to monitor or evaluate information provided by working groups, such as recommendations or the agencies' responses. FCC is also supporting the development of a new advisory committee, the Communications Security, Reliability, and Interoperability Council. FCC filed the charter for the council with the appropriate House and Senate committees in April 2007.[84] The charter did not state if or how FCC will monitor, evaluate, or respond to the recommendations made by the council.[85]

CONCLUSIONS

For the first time, the National Emergency Communications Plan provides an overarching strategy for emergency communications at all levels of government. This plan and other significant federal efforts represent an increasingly strategic approach by the federal government to enhance emergency communications and address existing vulnerabilities. Collaboration and monitoring remain critical components for success given the complex nature of emergency communications, the number of stakeholders involved, and the numerous efforts underway. Those federal agencies that do not use collaboration best practices jeopardize the success of not only their own efforts, but those of other agencies who have a role in supporting and enhancing emergency communications. Identified emergency communications vulnerabilities may not only persist, but deteriorate further as supporting infrastructure ages and technology continues to change. Establishing an effective governance structure by completing a MOU and establishing the Emergency Communications

Preparedness Center would improve the implementation of efforts that depend on the participation of multiple agencies, such as the National Emergency Communications Plan. Other federal efforts, such as FCC's 700 MHz Public/Private Partnership proceeding would also benefit from DHS and FCC establishing a common vision and mutually reinforcing strategies. This would help the agencies speak with one voice as they work with state, local, tribal, and private stakeholders. Given DHS's past experience and expertise in leveraging resources to assist states with emergency communications planning, it is well suited to offer similar assistance to federal agencies. Such assistance would help ensure that agencies plan for an emergency response, including evaluating how their communications assets and capabilities could best assist state and local first responders in disaster. Like collaboration, monitoring is crucial for ensuring advancement of federal efforts to enhance emergency communication. Improved monitoring and accountability of stakeholder and advisory committees recommendations—including agencies considering, deciding, and acting on such recommendations—would boost the value of these groups by monitoring agency responses, avoiding duplication of efforts, and identifying opportunities to work with other agencies.

Ultimately, the success or failure of federal efforts to enhance emergency communications will have the greatest effect on state and local first responders. Vulnerabilities involving continuity of communications, capacity, and interoperability can all cause communications failures during catastrophic disasters. As in the past, when future catastrophic disasters cause similar failures, the federal government will play a vital role in response. Addressing vulnerabilities through successful collaboration and monitoring of the wide variety of ongoing federal efforts will be essential in determining the quality of this future federal assistance to overwhelmed state and local first responders.

RECOMMENDATIONS FOR EXECUTIVE ACTION

We make four recommendations in this report to improve federal agencies' collaboration and monitoring in efforts related to emergency communications.

To help foster implementation of the National Emergency Communications Plan, we recommend that the Secretary of Homeland Security, in DHS's role as chair of the agency working group to establish the Emergency Communications Preparedness Center, work to complete the MOU to establish the center. The MOU should include a clear purpose, expected outputs, and realistic performance measures from participating agencies.

To help ensure that DHS and FCC's significant emergency communications efforts, such as the National Emergency Communications Plan and the 700 MHz Public/Private Partnership, have a common vision and mutually reinforcing strategies, we recommend that the Secretary of Homeland Security and the Chair of the Federal Communications Commission establish a forum, or other mechanism, to better collaborate on each agency's emergency communications efforts. Such collaboration could identify opportunities for aligning agency activities to ensure that they are mutually reinforcing, as well as developing an action plan or other working document to develop a common vision for implementation of the National Emergency Communications Plan and its relationship to the future 700 MHz Public/Private Partnership.

To help ensure that federal agencies and their communications assets are well-positioned to support state and local first responders in catastrophic disasters, we recommend that the Secretary of Homeland Security provide guidance and technical assistance to federal agencies in developing formal emergency communications plans. These plans could include identifying how federal agencies' communications resources and assets will support state and local first responders in a disaster.

To help DHS and FCC enhance the value of stakeholder groups' recommendations, we recommend that the Secretary of Homeland Security and the Chair of the Federal Communications Commission systematically track, assess, and respond to stakeholder groups' recommendations, including identifying actions taken by the agencies in response to recommendations, whether recommendations are duplicative with past recommendations, and opportunities to work with other agencies, as appropriate, to advance recommendations.

AGENCY COMMENTS

We provided a draft of this report to DHS, FCC, Commerce, Interior, and DOJ for official review and comment. In its comments, DHS generally agreed with our recommendations and noted that steps were already underway to implement some recommendations. Regarding our recommendation that DHS work to complete the Emergency Communications Preparedness Center's MOU, DHS stated that the agency had signed the MOU and that it had been circulated to the other interagency partners. While this represents progress, more work remains to complete the MOU and reach consensus among agencies on its purpose, expected outputs, and performance measures. Regarding our recommendation that DHS and FCC establish a mechanism for better collaboration on emergency communication efforts, DHS stated that its Office of Emergency Communications had begun regular coordination meetings with FCC's Public Safety and Homeland Security Bureau to identify areas for collaboration and to work jointly on common solutions as appropriate. Regarding our recommendation on DHS providing guidance and technical assistance to federal agencies in developing formal emergency communications plans, DHS noted that the Homeland Security Act focuses on assistance to "state, regional, local, and tribal governments," providing limited authority for DHS to provide assistance to other federal agencies. We agree that DHS cannot compel agencies to work to develop formal emergency communications plans, but this recommendation would include DHS offering such assistance through guidance and making available its expertise to other agencies. Regarding our recommendation that DHS systematically track, assess, and respond to stakeholder groups' recommendations, DHS noted that its Office of Emergency Communications has worked closely with numerous stakeholder groups to track and use the information from these groups within its other emergency communications efforts, such as the development of the National Emergency Communications Plan. We agree that DHS has some measures in place to track, assess, and respond to some stakeholder groups, but our recommendation calls for such a process to be applied systematically to all stakeholder group recommendations.

FCC generally agreed with our recommendations and provided comments via e-mail that we summarize here. FCC said that most aspects of the recommendations are already being actively pursued by the FCC, DHS and other federal agencies. In addition, FCC said that it was engaged in a large amount of work that goes beyond the report's recommendations aimed at improving federal responses and eliminating vulnerabilities not addressed in the report. In its comments, FCC said that the report relies heavily on anecdotal information and opinion, which are often uncritically presented as representing objective truth. For example, FCC questioned the use of several interviews with state and local officials about communications vulnerabilities forming the basis for much of the discussion of vulnerabilities in the report. As presented in our objectives, scope, and methodology, our case study work and related interviews were one component of identifying vulnerabilities. We also conducted a literature review of prior GAO products and other agency reports on emergency communications to ascertain and analyze common vulnerabilities.

Furthermore, the three primary vulnerabilities that we identified are similar to vulnerabilities identified by DHS, FCC, and other stakeholder groups.

FCC said that the report also lacks a sufficient number of facts about vulnerabilities, meaning that there will be no adequate way to judge whether the adoption of the recommendations actually improves emergency communications. Furthermore, FCC said that the report identifies vulnerabilities, but does not place them in context or suggest priorities in terms of how they should be addressed. As stated in our report, our recommendations will help federal efforts in addressing challenges to improve emergency communications by helping foster implementation of the National Emergency Communications Plan, helping ensure that significant emergency communications efforts share a common vision and have mutually reinforcing strategies, helping ensure that federal agencies and their communications assets are well-positioned to support state and local first responders, and by helping DHS and FCC enhance the value of stakeholder groups' recommendations. In addition, ranking, prioritizing, or suggesting how to address the vulnerabilities, was outside the scope of our work. We collected information from local and state emergency managers, law enforcement, firefighters, and other first responders, as well as federal officials and telecommunications industry officials, on efforts to address some of these vulnerabilities. We include this information on what jurisdictions are doing to address vulnerabilities throughout our report. Furthermore, our identification of vulnerabilities does not preclude FCC or another organization from exploring metrics or other benchmarks to track progress in addressing these vulnerabilities.

FCC also said that it was unclear whether the report is intended to address only first responder communications or all types of emergency communications, including commercial communications. We have clarified in our report that unless otherwise noted, when we refer to emergency communications systems, we mean those systems used by first responders. FCC also said that the report overlooks many vital issues, such as all the collaborative work done by federal agencies and communications companies to prepare and respond to communications disasters. We disagree that we omitted all collaborative work done by federal agencies and communications companies. We acknowledge in the report that DHS and other federal agencies have recently taken significant and strategic steps to enhance emergency communications and that a range of other federal efforts are underway. Additionally, we interviewed telecommunications industry officials as part of our audit work, and reported on the communications assets that companies can provide. Furthermore, we reported that private stakeholders, such as telecommunications companies and equipment manufacturers, have invested heavily to develop innovative technological solutions and expand or strengthen their networks for emergency responders and commercial use. We did not intend our report to include or highlight all technological capabilities present in certain emergency communications systems. We made changes to clarify the scope of our work, but remain confident about our findings and conclusions.

In its comments, Commerce provided information on two of its agencies'—NOAA and NTIA—roles in emergency communications. Commerce commented that emergency communications are important and that federal agencies must effectively coordinate to mitigate vulnerabilities. In its comments, Interior said that the report could have been improved if it incorporated Interior or federal interoperability collaboration efforts in regards to emergency response capabilities based on the following. First, we could have expanded the report's geographic scope such as including a case study involving fire. We do not intend to understate fire hazards by not including a fire scenario. However, our case studies represent a variety of different disaster scenarios representing different regions of the United States and we do not imply that these are the only possible catastrophic disaster scenarios. Interior said that we also could have conducted interviews with Interior's Emergency Management Offices. While we report on several federal efforts involving emergency communications, our examples do not constitute a complete list, or evaluation of the effectiveness of federal assistance currently available to first responders. Interior also said that the team could have reviewed existing emergency deployment systems capabilities and nationally recognized emergency and day-to-day interoperability efforts throughout the United States. Our work included conducting interviews with first responders and federal officials across the country and receiving information on communications capabilities and efforts to improve interoperability, among other things. DOJ did not provide comments on our draft report. DHS, FCC, Commerce, and Interior also provided technical comments that we then incorporated, where appropriate. DHS's, Commerce's, and Interior's letters are contained in appendices V, VI, and VII respectively.

As agreed with your offices, unless you publicly announce the contents of this report earlier, we plan no further distribution until 30 days from the report date. At that time, we will send copies of this report to the Secretary of Homeland Security, the Chair of FCC, the Secretary of Commerce, the Attorney General, the Secretary of the Interior, and appropriate congressional committees

Sincerely yours,

David J. Wise
Director, Physical Infrastructure Issues

Appendix I: Objectives, Scope, and Methodology

The objective of this report is to provide information on the status of emergency communications used by first responders. In particular, we sought to identify (1) vulnerabilities, if any, to emergency communication systems; (2) federal assistance available or planned to first responders for addressing any vulnerabilities or enhancing emergency communications; and (3) challenges, if any, with federal emergency communications efforts.

To identify vulnerabilities, if any, to emergency communication systems, we developed six case studies and subsequent analyses of varying catastrophic disaster scenarios both natural and man-made. In its National Response Framework, the Department of Homeland Security (DHS) defines catastrophic disasters as any natural or man-made incident that results in extraordinary levels of mass casualties, damage, or disruption severely affecting the population, infrastructure, environment, economy, national morale, and/or government functions.[86] Further, GAO has defined catastrophic disasters as a disaster whose effects almost immediately overwhelm the response capacities of affected state and local first responders and require outside action and support from the federal government and other entities.[87] The scenarios captured by our case studies contain elements from both of these definitions of catastrophic disasters.

Our case studies included a flood in northern California, a hurricane in southern Florida, a tsunami in Hawaii, a terrorist attack in Massachusetts, an earthquake in Tennessee, and a volcanic mudflow in the state of Washington. With this selection, we do not mean to imply that these are the only possible catastrophic disaster scenarios. The first step in selecting our particular case studies was to identify an exhaustive list of disaster scenarios facing communities across the United States. We limited this search to states, including Alaska and Hawaii. We conferred with subject matter experts, and reviewed data and documents from sources such as National Oceanic and Atmospheric Administration (NOAA), United States Geological Survey (USGS), and nongovernmental entities to produce a preliminary list of potential case studies. After producing our initial list of over 60 potential scenarios, we compared this list to available Geographic Information Systems (GIS) data, which ranked metropolitan areas by the types of hazards they face (i.e., seismic hazards, historic hurricane strikes, etc.). This GIS data included historical disasters from 1980 to 2006.[88] We also consulted USGS, NOAA, and other documents on historically less frequent disasters that could occur in our lifetimes. We identified overlap with our initial list and formulated several criteria to help select our final set of case studies. Our criteria included:

- National level impact—We selected scenarios that would have far-reaching impacts beyond the immediate location of the disaster. Responding to such disasters would be beyond the capacity of state and local officials, requiring assistance from the federal government.
- Likelihood of occurrence—We selected scenarios that had a higher likelihood of occurrence within our lifetime.
- Potential fatalities and injuries—We considered the population of scenario areas, as well as related casualty forecasts, models, and expert opinions to select scenarios that were more likely to cause higher numbers of fatalities and injuries.

- Economic impact—We considered potential economic losses, including damage to public and private infrastructure and the loss of public and private revenue.
- Diversity of catastrophic disaster—We selected a variety of scenario types to show how different disasters (i.e., earthquakes, hurricanes, tsunamis, and volcanic mudflows) may pose similar and/or different challenges to emergency communications systems.
- Diversity of geography—We selected scenarios to represent different regions of the United States.

Data were not available to fully evaluate all scenarios under our criteria. Previous research has not calculated estimated economic or other financial losses for all of our scenarios beyond a broad range (i.e., in the tens of billions). Also, we included one of our case studies, a terrorist attack in Boston, given continued high interest in terrorism by Congress and first responders. This case study did not directly follow our selectio n method for the other five, although some of the same criteria—such as potential fatalities and injuries, economic impact, and diversity of geography—still apply. Other criteria, such as likelihood of occurrence, have less application regarding the location of another terrorist attack. We selected Boston, Massachusetts, as the location of our terrorism scenario because it is among the top 20 population centers in the United States and is located in the Northeast, a region not represented by any of our other case studies. Furthermore, two of the planes involved in the September 11, 2001, terrorist attacks flew out of Boston's Logan International Airport. To provide context for a catastrophic terrorist attack, we used the Homeland Security Council's Planning Scenario Document.[89] This document provid es 15 all-hazards planning scenarios for use in national, federal, state, and local homeland security preparedness activities. These scenarios are designed to be the foundational structure for the development of national preparedness standards. Because our criteria included physical damage to communications systems, we used the Council's scenario involving a nuclear detonation of a 10-kiloton improvised device. When meeting with state and local stakeho lders for this scenario, we described details of this catastrophic disaster.

In developing our case studies, we visited site locations and interviewed local and state emergency managers; law enforcement, firefighters, and other first responders; and regional federal officials to help identify emergency communications vulnerabilities. We discussed catastrophic scenarios particular to each case study location and toured federal, state, and local emergency facilities. First responders also demonstrated available emergency communications equipment. We provide additional information on the hazards associated with each case study in appendix II. We conducted summary analyses of interviews and other information that we collected on these site visits. In addition to our case studies, we also conducted a literature review of prior GAO products and other agency reports on emergency communications to ascertain and analyze common vulnerabilities. The three primary vulnerabilities to emergency communications that we identified are similar to vulnerabilities identified by DHS, the Federal C ommunications Commission (FCC), and other stakeholder groups.

To identify federal assistance available to first responders for emergency communications, we interviewed officials and reviewed program documents from a variety of federal agencies with communications responsibilities and efforts underway or planned. Our work focused on the availability of federal assistance, but did not include a comprehensive evaluation of the effectiveness of this assistance. We reviewed recent emergency communications strategic guidance and documents from agencies such as DHS, FCC, and DOJ. We also reviewed FCC's 700 MHz Public/Private Partnership proceeding. Our work included a review of key planning documents such as the National Emergency Communic ations Plan. We also reviewed provisions of the Post-Katrina Act to identify new efforts underway to meet the act's emergency communications requirements. To obtain information regarding emergency communications grants and funding available to state and local first responders, we interviewed FEMA officials in the Grant Programs Directorate, as well as DOJ officials involved with law enforcement grants involving funding for emergency communications. To identify technical support, initiatives, and assets available to first responders in advance of or during a catastrophic disaster, we interviewed DHS and DOJ officials, and reviewed our recent work on several agency efforts. We also gathered information on available federal assistance during our case study work, collecting examples of state and local first responders' experiences and perceptions of federal guidance, grants, and other efforts.

To identify and examine any challenges with federal efforts to enhanc e emergency communications, we consulted our past work on emergency communications, interagency collaboration, and federal government program management and performance. Based on this review, we selected several best practices in collaboration—including establishing a common goal, developing mutually reinforcing strategies, and leveraging resources—and for government accountability and program performance relevant to emergency communications. We collected and analyzed key federal agency documents, such as DHS's National Emergency Communications Plan, National Communications Capabilities Report, National Preparedness Guidelines, FCC's notices for proposed rulemaking in the 700 MHz Public/Private Partnership proceeding, and other publicly available documents, such as comments filed to FCC from public safety entities. We determined the extent of interagency collaboration with regard to some significant federal efforts by comparing these efforts as described within these key agency documents, interviews with federal officials, state and local responders, and others against our selected best practices in collaboration. To determine the extent of monitoring being conducted by federal agencies of stakeholder groups' recommendations, we analyzed recommendations issued over the past 5 years to identify those which were repeatedly made and relevant to the vulnerabilities we identified. Additionally, we interviewed federal agency officials on what steps their respective agencies had taken to collaborate and monitor federal efforts, such as the Emergency Communications Preparedness Center and the National Emergency Communications Plan. We also interviewed state and local first responders, professional and trade group representatives, and telecommunications industry o fficials and reviewed testimony provided by these groups before Congress and FCC to obtain their perspectives on challenges to federal efforts.

We conducted this performance audit from February 2008 to May 2009 in accordance with generally accepted government auditing standards. Tho se standards require that we plan and perform the audit to obtain sufficient and appropriate evidence to provide a reasonable basis for our findings and conclusions based on our audit objectives. We believe that the evidence obtained provides a reasonable basis for our findings and conclusions based on our audit objectives.

APPENDIX II: CASE STUDY DISASTER SCENARIOS

The following provides additional background information and context on the hazards associated with each of our case study locations. These case studies included a flood in northern California, a hurricane in southern Florida, a tsunami in Hawaii, a terrorist attack in Massachusetts, an earthquake in Tennessee, and a volcanic mudflow in the state of Washington. The scenarios are hypothetical and we include descriptions of potential impacts, including the geographic area and populations affected, event frequency, hazards descriptions, and maps depicting a particular hazard on a national scale.

Sacramento Flooding

Disaster Type

A large flood in the city of Sacramento, California, and the surrounding Central Valley.

Geographic Area and Populations Affected

The city of Sacramento, the California state capital, is located in the Central Valley, which encompasses the floodplains of two major rivers— the Sacramento and the San Joaquin—as well as additional rivers and tributaries that drain from the Sierra Nevada Mountains. The approximately 1.8 million residents of Sacramento and the surrounding Central Valley would feel the most significant and direct effects of flooding. However, widespread flooding would likely have effects on the entire state of California by disrupting the state government (Sacramento's Capitol Building is below the flood water level), utilities, and drinking water. The Sacramento area provides two-thirds of the drinking water to Southern California.

Event Frequency

The actual number of years between floods of any given size varies greatly. Severe floods can occur in successive or nearly successive years. Scientists observe how frequently different sizes of floods occur, and the average number of years between them, to determine the probability that a flood of any given size will be equaled or exceeded during any year.

- A 100-year flood has water levels high enough that there is a 1 percent chance of being equaled or exceeded in any given year.
- A 500-year flood has water levels high enough that there is a 0.2 percent chance of being equaled or exceeded in any given year. Since 1951, the Sacramento area has experienced five major floods (see Figure 16 for a national map of flood declarations since 1980).

Hazards Description

- **Extensive flood zones:** There are three different types of flood events in the Sacramento area: flash, riverine, and urban storm water.[90] Flash floods are localized and the result of extensive rainfall. Riverine floods occur when riverbeds overflow into the flood zone. Existing flood zones in Sacramento County are extensive. Urban storm water floods result when urban drainage systems cannot handle the quantity of runoff from rainfall.

- **Aging infrastructure might fail:** The Central Valley's aging flood-control system provides only limited protection as many of the system's levees were poorly built or placed on top of older foundations up to 150 years old. Several areas of the county are subject to flooding by the overtopping of rivers and creeks, levee failures, and the failure of urban drainage systems to accommodate large volumes of water during severe rainstorms.

- **Other natural disasters can cause flooding:** While heavy rains are a major factor contributing to flooding, a major earthquake in the Bay Area could also destroy levies, which would result in massive flooding in the Sacramento area.

Miami Hurricane

Disaster Type

Major hurricane striking southern Florida near the City of Miami.

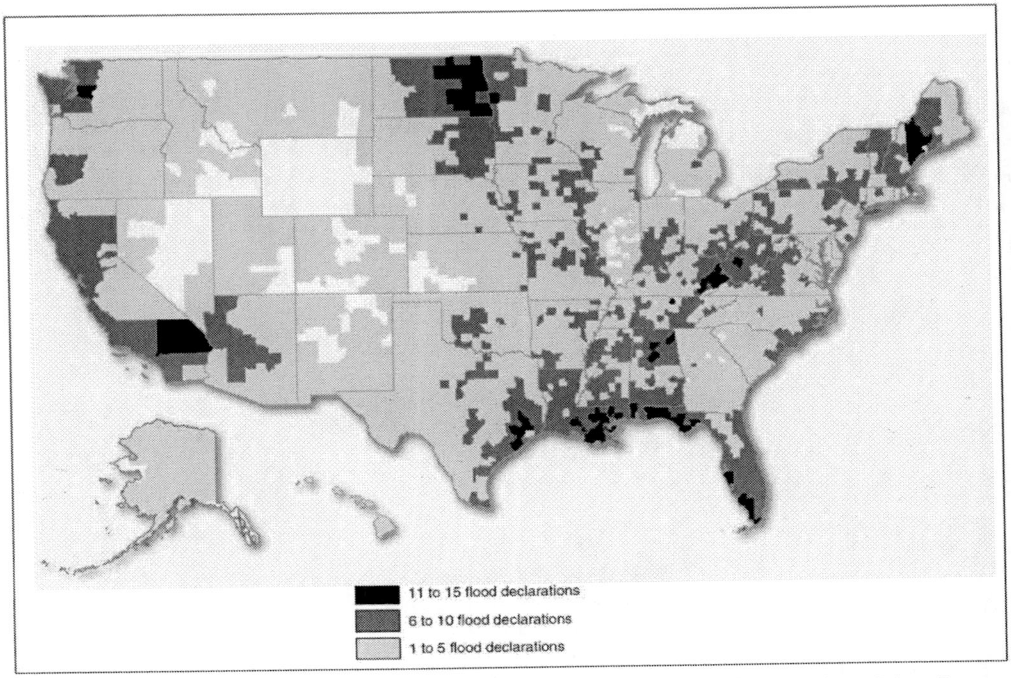

Sources: GAO analysis of Federal Emergency Management Agency (FEMA) data; Map Resources (map).

Figure 16. Number of Major Flood Declarations by County, 1980 – 2005.

Geographic Area and Populations Affected

A hurricane is a tropical cyclone storm system, which generally forms in waters with temperatures at or above 80 degrees Fahrenheit, such as those off of the U.S. coastlines in the Gulf of Mexico and Atlantic Ocean. A major hurricane striking the coast of Florida near the City of Miami and surrounding Miami-Dade County would bring high winds, heavy rains, and ecological damage to an area with approximately 2.4 million people. The effects of a major hurricane can include power and water outages, damage to buildings and roads, and restricted communication and rescue operations in the 24 hours following the storm.

Event Frequency

- "Hurricane season" is from June 1 through November 30. According to a NOAA official at the Tropical Prediction Center, August to October is the time of highest risk for hurricanes for southern Florida.
- Hurricane intensity is measured on the Saffir-Simpson hurricane scale, which classifies hurricanes on a scale of 1 to 5, based on the sustained wind speed. A category 1 hurricane has sustained winds of 74 to 95 miles per hour, while a category 5 has sustained winds greater than 155 miles per hour. A category 4 hurricane (storms with sustained winds of 131 to 155 miles per hour) or stronger hurricane hits southern Florida every 16 years. (See figure 17 for a national map showing the location and number of hurricane strikes, including southern Florida, since 1980.)

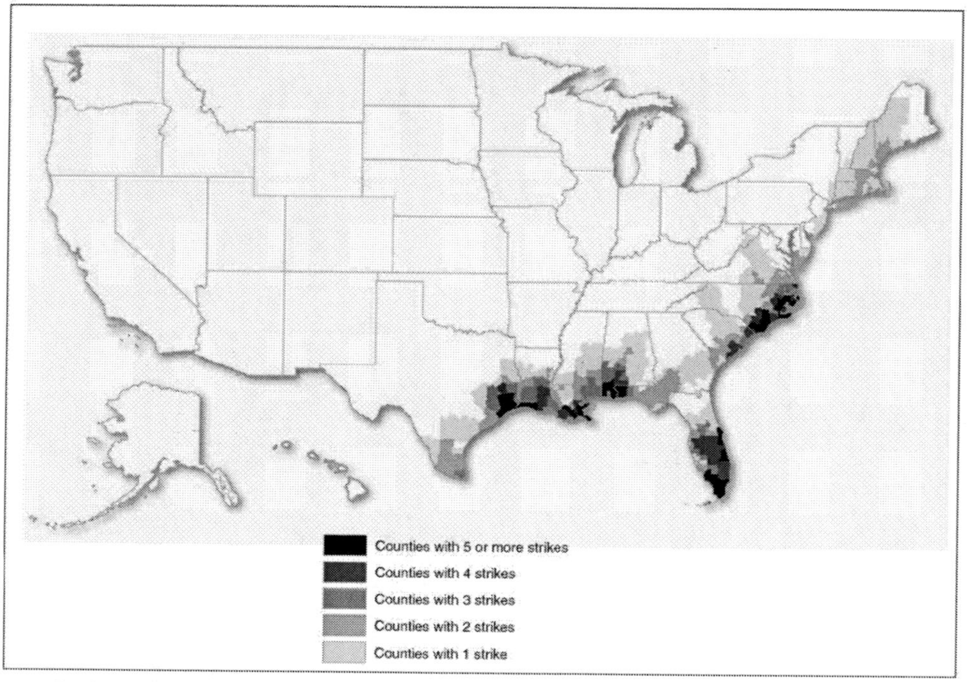

Sources: GAO analysis of National Oceanic and Atmospheric Administration (NOAA) data; Map Resources (map).

Figure 17. Number of Hurricane Strikes by County, 1980 – 2007.

Hazards Description

- **High Winds:** Hurricane-force winds can destroy constructed buildings and mobile homes. Debris such as signs, roofing material, and items left outside can become airborne in hurricanes. Downed trees and other debris, such as occurred when Hurricane Andrew struck southern Florida in 1992, would largely restrict movement, including aid, for the first few days. Local building codes account for and are designed to withstand high winds from moderately strong hurricanes, however, buildings are likely to suffer power and water outages, as well as have windows destroyed, potentially making them uninhabitable. A category 5 hurricane would test even these building codes, the most stringent in the nation for hurricanes.

- **Storm surge:** Water that is pushed toward the shore by the force of the winds swirling around the storm (the storm surge) combines with wave action and the normal tides to push water onshore to depths of as much as 15 feet or more. According to NOAA officials, these waters would penetrate inland at decreasing depths over eastern sections of Miami near the waterfront. They could extend even further inland up rivers.

- **Inland flooding:** According to NOAA officials, for the 35 years preceding Hurricane Katrina's landfall along the Gulf Coast, inland flooding was responsible for more than half the deaths associated with domestic hurricanes. While the City of Miami is less vulnerable than New Orleans (it is not situated below sea level), sections of the city have been flooded by rainfall associated with major hurricanes and weaker tropical cyclones.

- **Associated Tornadoes:** Hurricanes can also produce tornad oes, adding to the potential for destruction.

Honolulu/Hilo Tsunami

Disaster Type

A tsunami striking coastal communities in the state of Hawaii, including Hilo on the island of Hawaii and Honolulu on the island of Oahu.

Geographic Area and Populations Affected

Coastal communities in Hawaii are at high risk for tsunamis. Tsunamis potentially destructive to Hawaiian communities, including the cities of Hilo and Honolulu, may originate at distant locations around the perimeter of the Pacific Ocean, or may be locally generated by earthquakes on or near Hawaii. A tsunami originating in Alaska's Aleutian Islands would reach the Hawaiian Islands in 4.5 to 5.5 hours.

Event Frequency

- About 50 tsunamis have occurred in Hawaii since the early 1800s. Seven of these tsunamis caused major damage to Hawaii, and two of these tsunamis were locally generated near Hawaii. One of the most severe occurred in 1946 when a tsunami originating in the Aleutian Islands struck Hawaii without warning and killed over 170 people.[91]
- According to NOAA officials, Hawaii has a high risk for future tsunamis given its location in the middle of the Pacific Ocean, where about 80 percent of all recorded tsunamis have occurred.
- NOAA officials also reported that tsunamis hit Hawaii several times per century. It has been 34 years since the last tsunami struck Hawaii in 1975. (See figure 18 for a map of high and very high hazard coastal areas based on tsunami frequency.)

Hazards Description

- **Different triggers:** Tsunamis are large, rapidly moving ocean waves triggered by a major disturbance of the ocean floor—typically an earthquake—but sometimes by a sub-marine landslide or volcanic eruption. A tsunami wave can travel at speeds of 600 miles per hour.

- **Little to no warning:** Locally generated tsunamis are potentially the most dangerous, because the time between their generation and when the waves arrive on shorelines may be too brief to warn and evacuate people. After an earthquake near Hawaii in 1975, the first wave reached Hilo 20 minutes later. Distant tsunamis generally take hours to arrive, which would allow more time for evacuation.

- **Wave Damage:** As the tsunami approaches the coast, the wave speed slows as the wave height grows in the shallower waters, sometimes cresting at heights of 100 feet and striking the land at speeds of 30 mph or above. A series of waves may strike a coastline at intervals of every 5 to 40 minutes, and the first wave is often not the largest.[92] The size and destructiveness of the waves are largely determined by the local topography, both onshore and offshore, and the direction from which the wave approaches. A tsunami wave may be very small in the deep ocean, but can become a fast-moving wall of turbulent water as it approaches land.

Boston Terrorist Attack

Disaster Type

- Terrorist incident involving the detonation of a 10-kiloton improvised nuclear device in Boston, Massachusetts.[93]

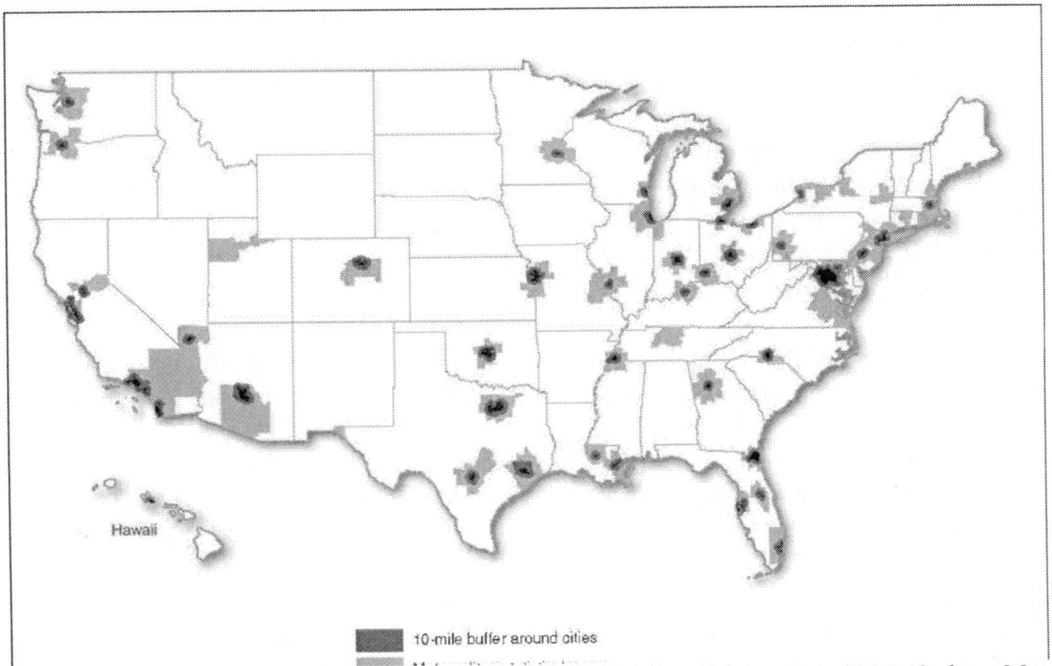

Sources: GAO analysis of National Oceanic and Atmospheric Administration (NOAA) data; Map Resources (map).

Figure 18. Tsunami Hazard Based on Frequency.

Geographic Area and Populations Affected

A nuclear bomb blast in a major metropolitan area such as Boston would cause widespread casualties, damage, and economic disruption. Approximately 600,000 people reside in the city of Boston, and over 3 million people live in the greater metropolitan area. The most severe effect of a 10-kiloton nuclear device would be felt within a few miles of the detonation point. Flying debris may damage areas within approximately 3.5 miles of the detonation point. Severe radiation fallout can cause acute health hazards up to 150 miles from point of detonation, and less severe radiation can cause contamination up to 3,000 miles from point of detonation.

Event Frequency

The likelihood of a terrorist attack is unknown; however, DHS has determined the Boston area to be at risk of attack and has designated it as an Urban Areas Security Initiative region. The criteria to determine the risk to urban areas includes and considers threats, vulnerabilities and consequences, such as threats from international terrorist networks and their affiliates (see figure 19 for a national map of these regions).[94]

Hazards Description

- **Detonation zone:** The intense heat of a nuclear explosion produces fires located throughout the immediate blast zone. Human casualties, damaged buildings, downed power and phone lines, leaking gas lines, broken water mains, and weakened bridges and tunnels are just some of the hazardous conditions that could result. If industrial storage facilities and manufacturing operations are located near the detonation site, additional hazardous materials could also be released.

- **Electro-magnetic pulse:** A nuclear explosion could also produce a high- voltage spike called an electro-magnetic pulse. This pulse radiates outwards from the detonation site and has the potential to disrupt the communications network, other electronic equipment, and associated systems within an approximately 3-mile range from the detonation point.

- **Damage to infrastructure:** There could be significant damage to general infrastructure, including transportation systems, power generation and distribution systems, communications systems, food distribution, and fuel storage and distribution. There could also be concerns about the safety and reliability of structures such as dams and hazardous material storage facilities. Structures that provide essential services, such as hospitals and schools, may also be damaged.

- **Radiation fallout:** The effects of the damage from the blast, radiation, and fallout could be significant within an approximately 3-mile range of the detonation point, with lesser effects on populations up to 3,000 miles away.

Sources: GAO analysis of Federal Emergency Management Administration (FEMA) data; Map Resources (map).

Figure 19. Urban Areas Security Initiative Regions, 2008.

Memphis Earthquake

Disaster Type

Major earthquake in the New Madrid seismic zone near Memphis, Tennessee.

Geographic Area and Populations Affected

- The New Madrid Seismic Zone is a collection of fault lines in the central United States. An earthquake in the New Madrid Seismic Zone earthquake could shake the entire Mississippi Valley, including the states of Tennessee, Missouri, Arkansas, Mississippi, Illinois, Kentucky, and Ohio. This area is home to millions of people and includes the cities of St. Louis, Missouri; and Memphis, Tennessee.

Event Frequency

- The zone has produced several major earthquakes since 1800 and geologists expect similar earthquakes in the future.[95] Geologists have dated evidence of past earthquakes at or exceeding 7 in magnitude to the years 900 and 1450.[96] This suggests that magnitude 7 or greater earthquakes reoccur in the region approximately every 500 years. The last series of 7 or greater earthquakes was in 1811-1812 (see figure 20 for a national map of earthquake hazards). The last earthquake over magnitude 6 in the New Madrid seismic zone was a 6.6 tremor in 1895.
- USGS has calculated the probability of a damaging earthquake of magnitude 6 or greater in the region to be between 25 to 40 percent in the next 50 years.

Hazards Description

- **Seismic waves:** Seismic wave energy decreases much more slowly in soils in the central and eastern United States than in western portions of the country. Consequently, there could be shaking over a larger area.

- **Unstable soil:** Muddy, sandy deposits found near rivers tend to liquefy during an earthquake, which causes buildings to sink, tip over, and otherwise destabilize.

- **Damage to infrastructure:** Potential losses from a major earthquake are expected to be significant due to buildings not designed and constructed to withstand strong shaking. A quake will most likely damage businesses, transportation, communication, oil and natural gas pipelines, and housing.

- **Economic losses:** Estimated building damage costs could run as high as $70 billion from one major earthquake alone. Economic costs from disruptions in commerce through the center of the country could cost additional billions.

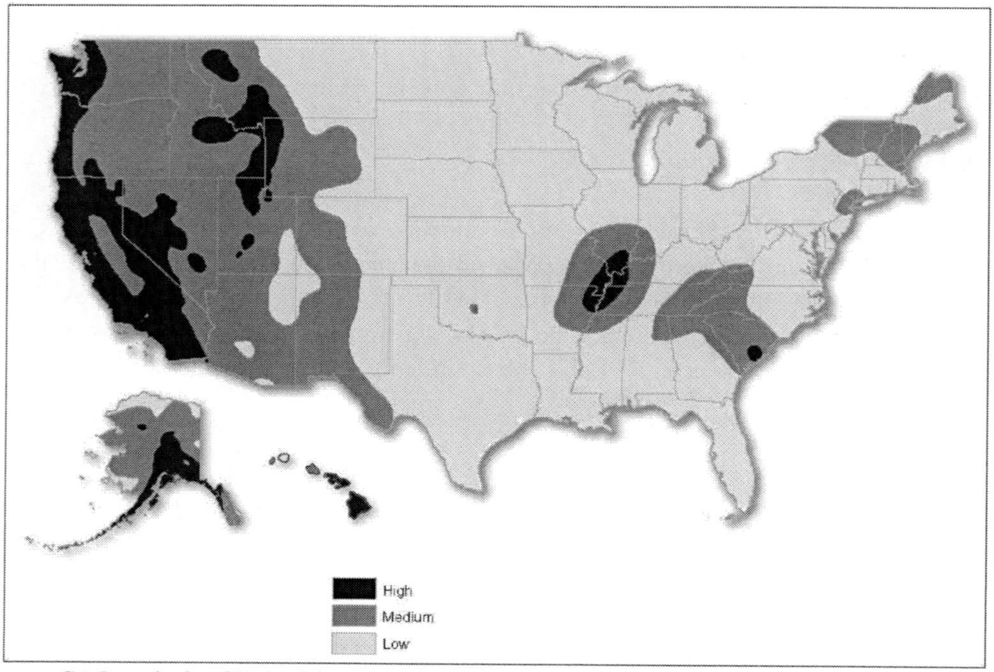

Sources: GAO analysis of United States Geological Survey (USGS) data; Map Resources (map).

Figure 20. High, Medium, and Low Seismic Hazards.

Mount Rainier Volcanic Mudflow

Disaster Type

A volcanic mudflow, also called a "lahar", descending from Mount Rainier and inundating communities in Washington state.

Geographic Area and Populations Affected

A major lahar originating from Mount Rainier in Washington state could inundate portions of the Puget Sound lowlands, including the towns of Orting, Puyallup, as well as portions of the city of Tacoma over 40 miles from the mountain's summit. Research indicates that Mount Rainier has been the source of many lahars that buried areas that are now densely populated.[97]

Event Frequency

- During the past few thousand years, lahars reaching the Puget Sound lowlands have occurred every 500 to 1,000 years. The last lahar to reach the Puget Sound lowlands occurred approximately 500 years ago. Past lahars have struck different areas in the vicinity of Mount Rainier. For example, during the last lahar to reach the Puget

Sound lowlands approximately 500 years ago, the lahar did not inundate the present location of the city of Tacoma.

- Smaller flows not extending as far as the Puget Sound lowlands occur more frequently. USGS estimates at least a one in seven chance of a lahar reaching the Puget Sound lowlands during an average human lifespan.
- USGS ranked Mount Rainier as a "very high threat volcano" among those volcanoes in the United States and its territories. (See Figure 21 for a national map of the location of USGS's high threat and very high threat volcanoes.) Of the 169 active volcanoes in the United States, USGS ranked 18 as very high threat volcanoes.

Hazards Description

- **Occur suddenly:** Lahars can occur with little or no warning. Most lahars large enough to flow beyond the boundaries of Mount Rainier National Park would occur during periods of volcanic unrest or eruption. For these large lahars, the estimated time between detection of a lahar on Mount Rainier and its arrival in the town of Orting, Washington, is about 40 minutes. Orting is over 10 miles from the boundary of Mount Rainier National Park and about 20 miles from the summit of Mount Rainier. Dispersed populations closer to Mount Rainier would be affected sooner.

- **Dangerous debris flow:** Lahars are fast-moving slurries of volcanic rock, mud, and water that look and behave like flowing concrete. Mount Rainier supports more than 1 cubic mile of glacial ice—as much as all other Cascade Range volcanoes combined. Thus, there is the potential to

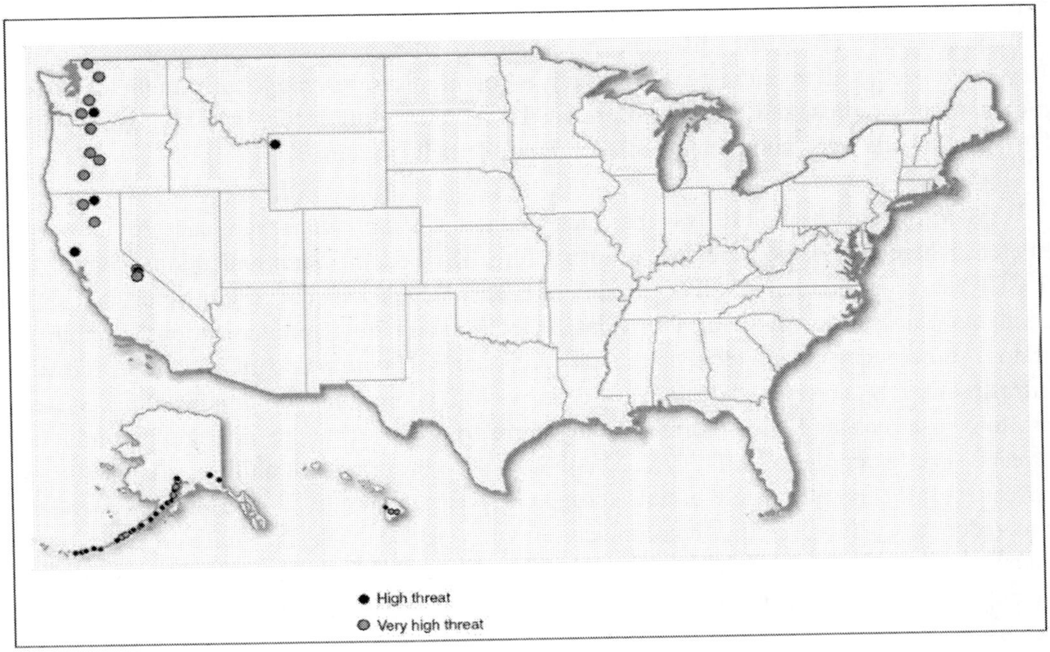

Figure 21. Location of High Threat and Very High Threat Volcanoes in the United States.

- **Different potential triggers:** Triggers for lahars do not have to be associated with volcanic eruptions. For example, a large flank collapse of the mountainside could also trigger a lahar at Mount Rainier. Although many flank collapses occur during eruptive periods, it is also possible for them to be triggered by earthquakes or result from the progressive weak ening of rock, saturation by groundwater, and the continuing pull of gravity.

APPENDIX III: DESCRIPTIONS OF COMMUNICATIONS SYSTEMS AND TECHNOLOGIES USED BY FIRST RESPONDERS

Land Mobile Radio System Land mobile radio systems are the primary means of communications among first responders. These systems typically consist of handheld portable radios, mobile radios, base stations, and repeaters. Land mobile radio networks operate on different spectrum frequencies, such as very high frequency (VHF), ultra high frequency (UHF), 700 MHz, and 800 MHz. FCC has reported that radio handsets must operate on the same frequencies to communicate. For example, a handset operating on a specific frequency in the UHF band will not be able to directly communicate with a handset operating on a different UHF frequency or on a VHF, 700 MHz, or 800 MHz frequency. Generally, first responders must carry multiple radios to allow direct communication with radio systems operating on different frequencies.

Handheld portable radios are typically carried by first responders and tend to have a limited transmission range. Mobile radios are often located in vehicles and use the vehicle's power supply and a larger antenna, providing a greater transmission range than handheld portable radios. Base station[98] radios are located in fixed positions, such as public service access points or dispatch centers, and tend to have the most powerful transmitters. A network is required to connect the different base stations to the same communications system. Repeaters are used to increase the effective communications range of handheld portable radios, mobile radios, and base station radios by retransmitting received radio signals. Figure 22 illustrates the basic components of a land mobile radio system.[99]

Satellite systems Satellite systems, such as phones, radio, and e-mail, can provide service in areas where there is no terrestrial infrastructure. The Federal Communications Commission (FCC) has reported that satellite communications, which can cover large portions of the Earth's surface, can provide an immediate backup emergency communications capability to restore emergency responder command and control communications when terrestrial infrastructure is severely damaged or destroyed.[100] Like other communications systems, orbiting satellites and their corresponding terrestrial infrastructure are not immune from threats. For example, satellites face unique space-based vulnerabilities. Typically, the terrestrial infrastructure, such as hub and gateway earth stations, is well protected, reliable, and redundant. Thus, satellite communications networks can weather terrestrial disasters if their associated earth stations survive, and can generally be restored to operation more quickly than terrestrial communications networks that rely on wireline infrastructure (see later discussion).

Cellular Systems First responders can use systems supported by cellular technologies, including cell phones. FCC has reported that cellular technologies, which offer "anytime, anywhere" mobility, could be an important tool for responders when their primary communications systems become unavailable. For example, first responders use cellular phones for non-critical primary communications or for backup communications when primary systems fail. The existence of multiple cellular service providers with national footprints greatly increases dependability and coverage even if individual commercial networks are suffering disruptions or do not necessarily meet all public safety requirements. If a cellular tower or its associated power is lost during a disaster, they could be temporarily replaced with a portable tower, backup generators, and other backup equipment.

Wireline Systems FCC has reported that first responders depend on wireline (landline) communications for operation of critical systems. Wireline service providers design networks to minimize single points of failure that could disrupt the network. However, the strategy of no single point of failure is not applied uniformly across the network. For reasons of economy, some systems' vulnerabilities may remain. In addition, facilities connecting first responders to central facilities may use copper cable, making them vulnerable to flooding, or they may use aerial cable, which subjects them to storm and fire damage. Loss of wireline facilities was well documented during Hurricane Katrina.

Technologies to Improve Interoperability

When different jurisdictions utilize different and incompatible systems, technologies such as audio switches, crossband repeaters, and others allow different systems to interoperate. These technologies to improve interoperability are described below.

Audio Switch An audio switch provides interoperability by sending an audio signal from one radio system to all other connected systems. An audio switch can be either stationary or mobile. One popular audio switch consists of a frame with slots, into which different hardware modules can be installed to control and interconnect different communications systems, such as VHF and UHF radios, as well as telephones. The audio switch can hold up to 12 interface modules, each capable of connecting a radio system. Audio switches are useful where multiple agencies temporarily come together to respond to an event because they are easily transportable and can be used to create temporary interoperability.

Crossband Repeater A crossband repeater provides interoperability between systems operating on different radio frequency bands by changing frequencies between two radio systems. Crossband repeaters can connect base stations or handheld or mobile radios. The repeater is also useful for extending the communications coverage beyond the range of a single radio. Crossband repeaters can also be linked together to overcome distances or geographical features blocking communication among users utilizing one repeater. *Console-to-console patch.* A console-to-console patch achieves interoperability by making an audio connection between the dispatch consoles of two different radio systems. Console-to-console patches connect consoles located at the dispatch centers where personnel receive incoming calls. These patches can connect personnel from an agency using one radio system to personnel from an agency using a different radio system. Connections between dispatch consoles can be made temporarily, as needed, through a public telephone line or permanently over a dedicated leased line or a dedicated microwave or fiber link.[101]

Software-defined Radios These radios use software to determine operating parameters such as the frequency band (such as VHF or UHF) and modulation type (such as AM or FM), and can be programmed to transmit and receive on any frequency within the limits of its hardware design. Software-defined radios will allow interoperability between agencies using different frequency bands, different operational modes (digital or analog), proprietary systems from different manufacturers, or different modulation (AM or FM). For example, a software-defined radio can be programmed to work as a conventional UHF radio but in another operating mode can function as an 800 MHz radio. Some software-defined radios could be used to identify unused frequencies and automatically make use of them, which is important in making efficient use of limited radio spectrum. The software-defined radio technology may also provide integrated voice and data over the same channel.

Voice over Internet Protocol Voice over Internet Protocol can connect different radio systems by using an Internet Protocol network as the connecting mechanism. Voice over Internet Protocol converts analog voice signals from a radio into digital data packets that travel over an Internet Protocol network.[102] At their destination, the digital information is converted back to analog audio and can be heard on the recipient's radio. Voice over Internet Protocol enables interoperability between agencies using different frequency bands, different operational modes (digital or analog), or proprietary systems from different manufacturers. Voice over Internet Protocol holds promise as a relatively low-cost solution to communications interoperability.

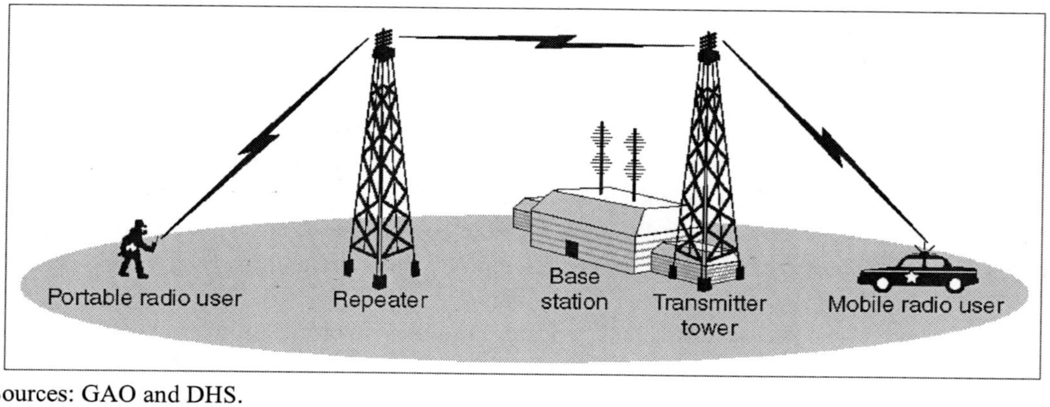

Sources: GAO and DHS.

Figure 22. Depiction of Land Mobile Radio System.

APPENDIX IV: STAKEHOLDER GROUP AND ADVISORY COMMITTEE DESCRIPTIONS

Name	Type of group	Mission	Year established
DHS Stakeholder Groups			
National Security Telecommunications Advisory Committee (NSTAC)	Stakeholder Group	To provide industry advice regarding national security and emergency preparedness and the availability and reliability of telecommunication services. Its goal is to develop recommendations to the President to assure vital telecommunications links through any event or crisis, and to help maintain a reliable, secure, and resilient national communications.	1982
SAFECOM Emergency Response Council	Stakeholder Group	To provide broad based input from the public safety community on its user needs to the SAFECOM program. A mechanism to share best practices, lessons learned, and guidance so that interested parties at all levels of government can learn from one another's experience, perspective, and expertise.	SAFECOM founded in 2001
Federal Partnership for Interoperable Communications	Stakeholder Group	To address federal wireless communi-cations interoperability by fostering intergovernmental cooperation. Coordinating body that focuses on technical and operational matters within the federal wireless communi-cations community, representing more than 40 federal entities.	1994
FCC Advisory Committees			
Network Reliability and Interoperability Council-VII	Advisory Committee	To partner with the FCC, the communications industry, and public safety to facilitate enhancement of emergency communications networks, homeland security, and best practices across the telecommunications industry.	2004
Independent Panel Reviewing the Impact of Hurricane Katrina on Communications Networks	Advisory Committee	To review the impact of Hurricane Katrina on the telecommunications and media infrastructure. The panel studied the impact of Hurricane Katrina on the telecommunications and media infrastructure, and made recommendations for improving disaster preparedness, network reliability, and communications among first responders.	2006
Joint Advisory Committee on Communications Capabilities of Emergency Medical and Public Health Care Facilities	Advisory Committee	To assess specific communications capabilities and needs of emergency medical and public health care facilities; options to accommodate growth of basic and emerging communications services; and options to improve integration of communications systems used by emergency medical and public health care facilities with existing or future emergency communications networks.	2007

Source: GAO analysis of DHS and FCC information.

Note: The FCC committees on this list no longer exist due to either termination or charter expiration. For example, the Network Reliability and Interoperability Council will be subsumed by the new Communications Security, Reliability, and Interoperability Council.

APPENDIX V: COMMENTS FROM THE DEPARTMENT OF HOMELAND SECURITY

U.S. Department of Homeland Security
Washington, DC 20528

Homeland Security

June 9, 2009

Mr. Dave J. Wise
Director
Physical Infrastructure Issues
Government Accountability Office
Washington D. C. 20548

Dear Mr. Wise:

Thank you for the opportunity to comment on the draft report: Emergency Communications: Vulnerabilities Remain and Limited Collaboration and Monitoring Hamper Federal Efforts (GAO-09-604). The Department of Homeland Security (DHS) appreciates GAO's work in planning, conducting, and issuing this report.

The following represents the DHS response to the recommendations contained in the draft report.

Recommendation:

The Secretary of Homeland Security, in DHS's role as chair of the agency working group to establish the Emergency Communications Preparedness Center(ECPC), work to complete the memorandum of understanding to establish the center.

Response: Concur: Office of Emergency Communication (OEC) concurs that finalizing the ECPC through the adoption of the Memorandum of Agreement (MOA) is an important step in formalizing the ECPC's role in coordinating emergency communications across member federal agencies and facilitating implementation of the NECP. While the final Memorandum of Understanding (MOU) has not yet been approved by all member agencies, it has been signed by DHS and circulated to interagency partners for final concurrence. Throughout the Charter development process, the ECPC has been actively working on a number of key coordination and NECP implementation issues by means of staff-level working groups, including the Grants Focus Group and the Technical Assistance Focus Group. The Grants Focus Group provides a forum for Federal grant issuers to convene to share lessons learned, best practices, and information on their specific grant programs further fostering alignment among the Federal communications grant programs. The Technical Assistance Focus Group aims to identify commonalities for Technical Assistance across the federal government.

Recommendation:

The Secretary of Homeland Security and the Chair of the Federal Communications Commission (FCC) establish a forum, or other mechanism, to better collaborate on each agency's emergency communications efforts.

- 2 -

Concur: OEC agrees that regular, senior-level meetings between DHS, NTIA and the FCC would be beneficial. OEC notes that OEC and the FCC's Public Safety and Homeland Security Bureau have begun regular coordination meetings to identify areas for collaboration and to work jointly on common solutions as appropriate. Additional
coordination will occur via the ECPC.

Recommendation:

The Secretary of Homeland Security provide guidance and technical assistance to federal agencies in developing formal emergency communications plans.

Concur: However, it should be noted that Title 18 of the Homeland Security Act focuses on assistance to "State, regional, local, and tribal governments," and provides limited authority for OEC to provide assistance to other federal agencies.

Recommendation:

The Secretary of Homeland Security and the Chair of FCC systematically track, assess, and respond to stakeholder groups' recommendations.

Concur: Though OEC concurs with this recommendation, it is important to note that OEC currently works closely with numerous stakeholder groups, including the SAFECOM Executive Committee (EC) and SAFECOM Emergency Response Council (ERC), as well as relevant critical infrastructure coordinating councils. EC/ERC meetings occur on a regular basis, and include the compilation and circulation of meeting notes and minutes. The Federal Partnership for Interoperability Coordination (FPIC) is another example of a stakeholder body that OEC has tracked and for which it has assessed recommendations. Indeed, in developing the NECP, OEC utilized these and other coordination forums to engage more than 150 stakeholders at all levels of government and the private sector. OEC also relied upon findings from the FCC's Hurricane Katrina Task Force, the FCC's Joint Advisory Committee on Public Health and the NSTAC Emergency Communications and Interoperability Task Report in developing the NECP.

Again, thank you and your staff for producing a thorough report.

Sincerely,

Jerald E. Levine
Director, Departmental GAO/OIG Liaison Office

Appendix VI: Comments from the Department of Commerce

UNITED STATES DEPARTMENT OF COMMERCE
The Secretary of Commerce
Washington, D.C. 20230

June 4, 2009

Mr. David J. Wise
Director, Physical Infrastructure Issues
Government Accountability Office
701 5th Avenue, Suite 2700
Seattle, WA 98104

Dear Mr. Wise:

Thank you for the opportunity to comment on the Government Accountability Office's (GAO) draft report entitled *Emergency Communications: Vulnerabilities Remain and Limited Collaboration and Monitoring Hamper Federal Efforts* (GAO-09-604). Technical and editorial comments to the draft report were provided to GAO staff earlier.

As part of this engagement, GAO met with officials from two of the Department of Commerce's agencies: the National Oceanic and Atmospheric Administration (NOAA) and the National Telecommunications and Information Administration (NTIA). Among other things, NOAA provides daily weather forecasts, severe storm warnings, and climate monitoring. Its dedicated scientists provide research and use high-tech instrumentation to provide citizens, planners, emergency managers, and other decisionmakers with reliable information that may be needed during a catastrophic or other emergency event. NTIA is responsible for managing the Federal Government's use of the radio frequency spectrum. NTIA also manages the Public Safety Interoperable Communications (PSIC) grant program, which awarded $968,385,000 to fund interoperable communications projects in 56 States and Territories.

In the draft report, GAO made recommendations to improve Federal agencies' collaboration and monitoring in efforts related to emergency communications. I agree that emergency communications among first responders are of the utmost importance and that Federal agencies must be able to effectively coordinate activities to mitigate vulnerabilities. Although none of the recommendations were directed at or required specific action on the part of NOAA or NTIA, I assure you that these agencies will continue to collaborate with and support other Federal agencies in reducing vulnerabilities related to emergency communications.

Thank you again for the opportunity to share the Department's comments on this draft report. The Obama Administration is committed to strengthening this country's preparedness, response, and recovery efforts.

Sincerely,

Gary Locke

APPENDIX VII: COMMENTS FROM THE DEPARTMENT OF THE INTERIOR

United States Department of the Interior
OFFICE OF THE SECRETARY
Washington, DC 20240

TAKE PRIDE IN AMERICA

JUN 3 2009

David Wise
Director, Physical Infrastructure Issues
U.S. Government Accountability Office
441 G Street, N.W.
Washington, D.C 20548

Dear Mr. Wise:

Thank you for providing the Department of the Interior the opportunity to review and comment on the draft Government Accountability Office Report entitled "EMERGENCY *COMMUNICATIONS: Vulnerabilities Remain and Limited Collaboration and Monitoring Hamper Federal Efforts*," (GAO-09-604).

We appreciate the diligent work of the team that prepared the report and the data collected. In general, we believe that this report is somewhat informative but could have been improved if it incorporated DOI or Federal interoperability collaboration efforts in regards to Emergency Response capabilities based upon the following:

- a perspective that was expanded geographically;
- interviews with DOI Emergency Management Offices;
- review of existing emergency deployment systems capabilities; and
- review of nationally recognized emergency and day-to-day interoperability efforts throughout the United States.

In addition, we suggest further interviews with the National Interagency Fire Center for a firsthand view of Emergency Communications support, capabilities and established federal practices.

The enclosure includes information regarding national policy and technical comments on the draft report.

If you have any questions, or need additional information, please contact Christopher Lewis at Christopher_Lewis@ios.doi.gov or phone, 703-648-5550.

Sincerely,

Pamela Haze

Pamela K. Haze
Deputy Assistant Secretary – Budget and Business Management

Enclosure

End Notes

[1] See GAO, Results-Oriented Government: Practices That Can Help Enhance and Sustain Collaboration among Federal Agencies, GAO-06-15 (Washington, D.C.: October 2005); and Executive Guide: Effectively Implementing the Government Performance and Results Act, GAO/GGD-96-118 (Washington, D.C.: June 1996).

[2] Homeland Security Council, Planning Scenarios – Executive Summaries, Version 2.0 (Washington, D.C.: July 2004).

[3] DHS's National Response Framework defines a catastrophic incident as any natural or man-made incident, including terrorism that results in extraordinary levels of mass casualties, damage, or disruption severely affecting the population, infrastructure, environment, economy, national morale, and/or government functions. A catastrophic incident could result in sustained nationwide impacts over a prolonged period of time; almost immediately exceeds resources normally available to state, tribal, local, and private- sector authorities in the impacted area; and significantly interrupts governmental operations and emergency services to such an extent that national security could be threatened. Department of Homeland Security, National Response Framework, 2008.

[4] GAO, Homeland Security: DHS Improved Its Risk-Based Grant Programs' Allocation and Management Methods, but Measuring Programs' Impact on National Capabilities Remains a Challenge, GAO-08-488T (Washington, D.C.: Mar. 11, 2008).

[5] 9-11 Commission, The 9-11 Commission Report: Final Report of the National Commission on Terrorist Attacks Upon the United States (Washington, D.C.: July 2004).

[6] Homeland Security Act of 2002, Pub. L. No. 107-296, 116 Stat. 2135 (2002).

[7] The Post-Katrina Act was enacted as Title VI of the Department of Homeland Security Appropriations Act, 2007, Pub. L. No. 109-295, 120 Stat. 1355 (2006). The provisions of the Post-Katrina Act became effective upon enactment, October 4, 2006, with the exception of certain organizational changes related to FEMA, most of which took effect on March 31, 2007.

[8] SAFECOM is a DHS communications program that provides research, development, testing and evaluation, guidance, tools, and templates on interoperable communications- related issues to local, tribal, state, and federal emergency response agencies.

[9] These policies include 9-1-1 and E9-1-1; operability and interoperability of public safety communications; communications infrastructure protection, and disaster response; and network security and reliability.

[10] Spectrum is divided into frequency bands, each having technical characteristics that affect electronic transmission in different ways. "Bandwidth" is related to the transmission capacity of a frequency band. If voice calls and low-rate data are involved, narrowband systems are adequate; but, to transmit video and images, broadband is needed.

[11] Omnibus Reconciliation Act of 1993, Pub. L. No. 103-66, § 6002, 107 Stat. 312, 387-392 (1993), codified as amended at 47 U.S.C. § 309(j).

[12] In September 2008, FCC issued its Third Further Notice of Proposed Rulemaking in this proceeding. See, Service Rules for the 698-746, 747-762 and 777-792 Bands; Implementing a Nationwide, Broadband, Interoperable Public Safety Network in the 700 MHz Band, Third Further Notice of Proposed Rulemaking, 23 FCC Rcd 14301 (2008) (700 MHz).

[13] DTV Delay Act, Pub. L. No. 111-4, 123 Stat. 112. The act also extends the Commission's auction authority through September 2012.

[14] In the Public/Private Partnership, the winning bidder of the commercial license in the Upper 700 MHz D Block (758-763/788-793 MHz) ("D Block") is to partner with the nationwide licensee of the public safety broadband spectrum (763-768/793-798 MHz) ("Public Safety Broadband Licensee") to enable construction of an interoperable broadband network that would serve both commercial and public safety users.

[15] NTIA is the President's principal adviser on telecommunications and information policy issues, and in this role frequently works with other Executive Branch agencies to develop and present the Administration's position on these issues.

[16] GAO, Radio Communications: Congressional Action Needed to Ensure Agencies Collaborate to Develop a Joint Solution, GAO-09-133 (Washington, D.C.: December 2008).

[17] GAO, Results-Oriented Government: Practices That Can Help Enhance and Sustain Collaboration among Federal Agencies, GAO-06-15 (Washington, D.C.: October 2005); and GAO, Executive Guide: Effectively Implementing the Government Performance and Results Act, GAO/GGD-96-118 (Washington, D.C.: June 1996).

[18] GAO/GGD-96-118 and GAO-06-15.

[19] FCC, however, would not necessarily characterize "continuity of communications" as a vulnerability; rather, FCC views this as a goal for emergency communications systems.

[20] While based on our case studies, we do not intend figure 4 or other figures depicting disaster scenarios and their effects on emergency communications to represent effects at any particular location that we visited. An actual

catastrophic disaster could have much larger and more complicated impacts. The figure is meant to provide examples of just some of the ways in which communications may be disrupted.

[21] A lahar is a volcanic mudflow that originates from the slopes of a volcano. These flows contain rock and other debris that exert high impact force against objects in their path, such as buildings and trees. Sizes vary, but lahars can travel over 50 miles from a volcano. Triggers for lahars include volcanic eruptions and massive landslides, such as the one that occurred at Mount St. Helens in the state of Washington in 1980.

[22] An emergency operations center is the physical location where multiagency coordination occurs. The core functions of such a center include coordination, communications, resource allocation and tracking, and information collection, analysis, and dissemination for disaster response.

[23] The New Madrid seismic zone is a collection of fault lines that runs through several states, including Arkansas, Missouri, Illinois, Kentucky, and Tennessee. The zone has produced several major earthquakes since 1800. Geologists expect similar earthquakes in the future.

[24] Established in 1983, the consortium's primary mission is to reduce deaths, injuries, property damage, and economic losses from earthquakes in the central United States. The organization's primary objective is to support multi-state response and recovery planning, resource acquisition; public education and awareness; promotion; mitigation, and research associated with earthquake preparedness in the central United States. Members include Alabama, Arkansas, Illinois, Indiana, Kentucky, Mississippi, Missouri, and Tennessee.

[25] These assets are mobile, self-contained cell sites to boost coverage for first responders. They are designed for short-term response and can process thousands of calls per hour.

[26] Underwater earthquakes typically generate tsunamis—landslides, volcanic activity, and meteor strikes are less common sources. Tsunami generating earthquakes usually occur in subduction zones, such as those found in the Pacific Ocean off the U.S. western and Alaskan coasts. Subduction zones are formed where one of the earth's outer shell of tectonic plates plunges underneath another. A tsunami's size depends on the earthquake's size, its depth below the ocean floor, the type and amount of seafloor movement and the energy released among other factors. Some tsunami waves can travel up to 600 miles-perhour, hitting nearby coasts within minutes and other distant shorelines hours later. We have previously reported on communications challenges related to tsunamis. See GAO, U.S. Tsunami Preparedness: Federal and State Partners Collaborate to Help Communities Reduce Potential Impacts, but Significant Challenges Remain, GAO-06-519 (Washington, D.C.: June 5, 2006).

[27] GAO, First Responders: Much Work Remains to Improve Communications Interoperabiity, GAO-07-301 (Washington, D.C.: Apr. 2, 2007).

[28] Spectrum allocations for state and local public safety are fragmented into many distinct slices of the radio spectrum. Bands of interest to public safety include VHF (very high frequency), and UHF (ultra high frequency). Radio systems used by law enforcement and other first responders operating in the 806-824 MHz and 85 1-869 MHz portion of the UHF bands are often referred to as "800 MHz" systems. The 800 MHz band is also home to commercial wireless carriers and private radio systems. In July 2004, the FCC adopted a comprehensive plan to reconfigure the 800 MHz band to separate public safety systems in the band from commercial wireless systems using cellular architecture.

[29] FCC officials noted limitations with patched networks. For example, handsets programmed to operate on frequencies not supported by base stations in the same area will still not be able to communicate with each other unless there is a compatible base station with which the handset can communicate.

[30] The 9-11 Commission Report: Final Report of the National Commission on Terrorist Attacks Upon the United States.

[31] Our examples do not constitute a complete list, or evaluation of the effectiveness of, this federal assistance currently available to first responders. We have previously reported on a number of issues/challenges in past disaster preparedness efforts. See GAO-09-133; GAO-07-301; GAO, *Homeland Security: Federal Leadership and Intergovernmental Cooperation Required to Achieve First Responder Interoperable Communications*, GAO-04-740 (Washington, D.C.: July 2004).

[32] Department of Homeland Security, *National Emergency Communications Plan* (Washington, D.C.: July 2008).

[33] Department of Homeland Security, *National Communications Capabilities Report* (Washington, D.C.: July 2008).

[34] The Emergency Communications Preparedness Center membership has since been broadened to include other federal agencies beyond those specified in the Post-Katrina Act.

[35] Interior is also a charter member of the Emergency Communications Preparedness Center, and the agency has participated in the Charter writing and review process.

[36] The D Block refers to the portion of commercially allocated spectrum that is adjacent to the public safety broadband spectrum. The March 2008 auction received only a single bid that did not meet the reserve price of $1.33 billion and thus did not become a winning bid.

[37] The Federal Advisory Committee Act, 5 U.S.C. App. 2, governed the operations of the Independent Panel Reviewing the Impact of Hurricane Katrina on Communications Networks with guiding principles of openness in government; diversity in membership and advice; and public accountability.

[38] The Post-Katrina Act directs these working groups to be established in each of the 10 FEMA regional offices and to include DHS, FCC, and other federal agencies with responsibility for coordinating interoperable emergency communications or providing emergency support services. 6 U.S.C. § 575.

[39] The Urban Areas Security Initiative is intended to enhance regional preparedness for prevention, protection, response, and recovery in 62 major metropolitan areas determined to be "highest risk."

[40] The State Homeland Security Initiative provides funds to states and territories to implement the goals and objectives of state homeland security strategies and initiatives included in the State Preparedness Report.

[41] The Interoperable Emergency Communications Grant Program provides governance, planning, training and exercise, and equipment funding to state, territories, and local and tribal governments to carry out initiatives to improve interoperable emergency communications. We provide more details on this program later in our report.

[42] The Emergency Operations Center Grant Program supports Emergency Operations Centers with a focus on addressing identified deficiencies and needs.

[43] The 25 Cities Project refers to the High-Risk Metropolitan Area Interoperability Assistance Project, a DOJ Wireless Management Office grant program that identified the top 25 metropolitan areas that were considered likely targets for terrorist attack and provided communication solutions to federal and local authorities such as fire, police, and emergency medical service s. Projects differed from city to city.

[44] The Community Oriented Policing Interoperable Communications Technology Program funded projects that used equipment and technologies to increase interoperability among the law enforcement, fire service, and emergency medical service communities from fiscal years 2003 through 2006.

[45] While we did not evaluate the effectiveness or quality of the Public Safety Interoperable Communications Grant Program as a part of our work, a recent Department of Commerce Office of Inspector General report identified an issue regarding Public Safety Interoperable Communications grant effectiveness. The report found that most grantees have made little progress in implementing their projects, and much remains to be done for the grantees to finish their projects by the September 30, 2010, deadline. See *Public Safety Interoperable Communications Grant Program: Grantees Appear Unlikely to Finish Projects Within Short Funding Time Frame*, Final Audit Report No. DEN-19003 (Washington, D.C.: March 2009).

[46] A tabletop exercise is a discussion-based exercise that focuses on existing plans, policies, mutual aid agreements, and procedures used among multiple agencies. Typically, a tabletop exercise involves representatives from the entire range of agencies and jurisdictions that would take action in all-hazards or terrorist response incidents.

[47] The Government Emergency Telecommunications Service uses a calling card that provides access authorization and priority treatment to first responders in the public switched telephone network through a unique dialing plan and personal identification number, and is designed to maximize all available telephone resources should outages occur during a disaster or other emergency.

[48] Project 25 was initiated in 1989.

[49] See GAO-07-301. Project 25 radios were marketed to and purchased by federal, state, and local agencies without any formal compliance testing to validate vendors' compliance with the standards.

[50] See S. Rep. No. 109-88, at 45 (2005); H.R. Rep. No. 109-241, at 81 (2005).

[51] The initial Compliance Assessment Program process began in December 2008, and after a 6-month grace period, equipment covered by the program that is purchased with federal grant dollars will be accompanied by declarations of compliance and test reports.

[52] The mission of FCC's Public Safety and Homeland Security Bureau is to collaborate with others, including other federal agencies. Specifically, the mission is "To collaborate with the public safety community, industry, and other government entities to license, facilitate, restore and recover communications services used by the citizens of the United States, including first responders, before, during and after emergencies by disseminating critical information to the public and by implementing the Commission's policy initiatives."

[53] GAO/GGD-96-118 and GAO, *Electronic Government: Potential Exists for Enhancing Collaboration on Four Initiatives*, GAO-04-6 (Washington, D.C.: October 2003).

[54] GAO-09-133. See also, GAO-06-15, GAO/GGD-96-118, and GAO-04-6.

[55] Similarly in 2004, GAO reported that SAFECOM's authority and ability to oversee and coordinate federal and state efforts for increased interoperability was limited by its dependence upon other agencies for funding and their willingness to cooperate. GAO, *Project SAFECOM: Key Cross-Agency Emergency Communications Effort Requires Stronger Collaboration*, GAO-04-494 (Washington, D.C.: April 2004).

[56] Collaboration can be broadly defined as any joint activity that is intended to produce more public value than could be produced when the organizations act alone.

[57] GAO-09-133.

[58] GAO-07-301.

[59] GAO-04-6.

[60] *Interoperability in the Next Administration*: *Assessing the Derailed 700 MHz D Block Public Safety Spectrum Auction*: Hearing Before the House Subcommittee on Emergency Communications, Preparedness, and Response, 110th Cong. (Sept. 16, 2008).

[61] Congressional members have expressed interest in the relationship between DHS, the Emergency Communications Preparedness Center, the National Emergency Communications, and the 700 MHz Public/Private partnership proceeding.

[62] FCC has taken steps to leverage DHS expertise on other efforts such as the 700 MHz narrowband requirement for mandatory interoperability through the use of Project 25 standards, as well as designated interoperability channels.

[63] FCC officials strongly assert that FCC must take into account input from all stakeholders, not just the views of DHS.

[64] According to DHS officials, while the National Emergency Communications Plan does not reflect the FCC's final policies for the 700 MHz Public/Private partnership, one of the key objectives of the plan (Objective 4: Standards and Emerging Communications Technologies) identifies a number of initiatives to ensure that emerging technologies, such as the wireless broadband technologies in the 700 MHz spectrum, are fully integrated with current emergency communications capabilities and work to improve interoperability on a nationwide basis.

[65] At a congressional hearing, members expressed interest in DHS officials providing their opinion on how FCC's 700 MHz proceeding should be structured.

[66] NTIA, acting on behalf of interested federal agencies, did submit formal comments on this proceeding. As previously discussed, NTIA has a role in managing spectrum used by federal agencies. Accordingly, discussions on the 700 MHz Public/Private Partnership that deal specifically with spectrum issues and access for federal agencies would include NTIA.

[67] The Congressional Research Service (CRS) has also highlighted the separate paths that DHS and FCC are pursuing in developing a national capabilities approach. Specifically, CRS stated that according to testimony "neither agency has undertaken to incorporate each other's goals in their specific planning." U. S. Congressional Research Service, *Public- Private Partnership for a Public Safety Network: Governance and Policy* (RL 34054, Oct. 16, 2008) by Linda K. Moore.

[68] At a congressional hearing, members expressed interest in the relationship between FCC's 700 MHz Public/Private Partnership and the Regional Emergency Communications Coordination Working Groups.

[69] This is not intended to be an exhaustive list or proposal for how the 700 MHz Public/ Private Partnership network could or should be aligned with the National Emergency Communications Plan or other DHS efforts. We also recognize that FCC has sought comment on aspects of its proposal and that a final order has not been issued. However, we believe that steps could be taken, as appropriate, to ensure that FCC's 700 MHZ Public Private Partnership plan, as adopted and implemented, is supportive of the National Emergency Communications Plan

[70] FCC officials noted that it was ineffectual to compare the strategic goals of the National Emergency Communications Plan to the performance benchmarks and build out requirements in the 700 MHz Public/Private Partnership proceeding. According to these officials, the plan is focused on improving the interoperability of current systems, while FCC's initiative would create a 100 percent interoperable network once built. Although we recognize the different emphasis, we continue to believe that FCC build out requirements could be leveraged to help meet the goals articulated in the National Emergency Communications Plan.

[71] According to FCC, it sought comments on a state role in coordinating participation in the network and received conflicting comments on these proposals.

[72] In the Third Further Notice, FCC tentatively concluded that it would reaffirm its prior decision that it was the "sole discretion" of the public safety broadband licensee whether to permit federal public safety agency use of the public safety broadband spectrum. This decision was supported by NTIA in its comments filed to FCC.

[73] FCC mandated that the security and encryption be consistent with state of-the-art technology. However, because FCC did not expressly reference the Advanced Encryption Standard, it is unclear whether FCC's mandated encryption will adhere to the standard called for in the National Emergency Communications Plan.

[74] *Public Hearing on Public Safety Interoperable Communications – The 700 MHz Band Proceeding*: Hearing Before the Federal Communications Commission (July 30, 2008). http://www.fcc.gov/realaudio/mt073008.ram, accessed April 2009.

[75] It was beyond the scope of work for this engagement to assess the full extent of emergency communications planning completed by federal agencies. Not all federal agencies have communications infrastructure or the role/responsibility to assist state and local first responders with communications.

[76] U.S. Executive Office of the President, *The Federal Response to Hurricane Katrina: Lessons Learned* (Washington, D.C.: February 2006).

[77] GAO/GGD-96-118.

[78] We have previously reported on the need for standard channel nomenclature to facilitate interoperability. GAO, *Homeland Security; Federal Leadership Needed to Facilitate Interoperable Communications among First Responders*, GAO-04-1057T (Washington, D.C.: September 2004).

[79] According to FCC officials, the agency is under no obligation to adopt any particular stakeholder group recommendation. Under the Federal Advisory Committee Act, the function of an advisory committee is to

provide advice; a federal agency may implement or not implement an advisory committee's recommendation at its discretion.

[80] We have previously reported on the composition and transparency of federal advisory committee selection processes. GAO, *Issues Related to the Independence and Balance of Advisory Committees,* GAO-08-611T (Washington, D.C.: April 2008).

[81] GAO, *Federal Advisory Committees Follow Requirements, but FCC Should Improve Its Process for Appointing Committee Members*, GAO-05-36 (Washington, D.C.: December 2004).

[82] *Recommendations of the Independent Panel Reviewing the Impact of Hurricane Katrina on Communications Networks*, Notice of Proposed Rulemaking, 21 FCC Rcd 7320 (2006).

[83] *Recommendations of the Independent Panel Reviewing the Impact of Hurricane on Communications Networks,* Order, 22 FCC Rcd 10541 (2007).

[84] Although the Communications Security, Reliability, and Interoperability Council became technically operational the date the charter was filed, members were never determined and no meetings have been held. The charter was renewed on March 19, 2009, and the FCC recently issued a Public Notice seeking nominations for membership on the committee. According to FCC officials, the Network Reliability and Interoperability Council will be subsumed by the new Communications Security, Reliability, and Interoperability Council.

[85] The purpose of an advisory committee charter is to describe the mission, goals, and objectives of the advisory committee (41 CFR 102-3.75), and according to the FCC, the Communications Security, Reliability, and Interoperability Council charter fulfilled the relevant legal requirements.

[86] Department of Homeland Security, *National Response Framework* (Washington, D.C.: January 2008).

[87] GAO, *Catastrophic Disasters: Enhanced Leadership, Capabilities, and Accountability Controls Will Improve the Effectiveness of the Nation's Preparedness, Response, and Recovery System*, GAO-06-618 (Washington, D.C.: Sept. 6, 2006).

[88] The GIS data did not include tsunami hazards. We consulted other data and included Honolulu, Hawaii (as well as other population centers in Hawaii) for several reasons. These included the example of the catastrophic tsunami that hit Indonesia in 2005, tsunamis repeatedly striking Hawaii over the past century, and the unique challenge of Hawaii's relative geographic isolation to the U.S. mainland.

[89] Homeland Security Council, *Planning Scenarios – Executive Summaries*, Version 2.0 (Washington, D.C.: July 2004). We also used this document to provide context for some of our other scenarios, such as the effects of major hurricanes and earthquakes.

[90] Sacramento County Hazard Identification, Multi-Hazard Mitigation Plan (December 2004).

[91] Following the 1946 tsunami, a tsunami warning system for the Pacific basin was developed. Presently, the Pacific Tsunami Warning Center System, which has its headquarters in Honolulu, is administered by the National Weather Service under NOAA.

[91] *U.S. Tsunami Preparedness: Federal and State Partners Collaborate to Help Communities Reduce Potential Impacts, but Significant Challenges Remain*, GAO-06-519 (Washington D.C.: June 2006).

[93] Homeland Security Council, *Planning Scenarios – Executive Summaries, Version 2.0* (Washington, D.C.: July 2004).

[94] *Homeland Security: DHS Risk-Based Grant Methodology Is Reasonable, But Current Version's Measure of Vulnerability Is Limited*, GAO-08-852 (Washington D.C.: June 2008).

[95] U.S. Geological Survey Fact Sheet FS-131-02 (October 2002).

[96] Earthquake magnitude is a measure of the size of an earthquake and is based on ground motions recorded on seismographs.

[97] U.S. Department of the Interior, U.S. Geological Survey, Fact Sheet-034-02, *Mt. Rainier – Learning to Live with Volcanic Risk* (2002).

[98] A base station contains the equipment for transmitting and receiving the radio signals that allow portable radios to communicate with each other.

[99] GAO-07-301, p. 11.

[100] Federal Communications Commission, *FCC Report to Congress: Vulnerability Assessment and Feasibility of Creating a Back-Up Emergency Communications System, Submitted Pursuant to Public Law No. 110-53* (Washington, D.C.: Jan. 30, 2008).

[101] A leased line refers to a permanent telephone connection set up by a telecommunications provider between two geographic locations. A fiber link uses light sent over a glass or plastic fiber to carry communication signals. A microwave link uses radio beams of extremely high frequencies to send information between two fixed geographic sites.

[102] In some cases, this is the Internet; in others, it is a private data network.

INDEX

A

accountability, 151, 165, 167, 173, 193
adjustment, 19
age, 22, 106, 107
algorithm, 5, 19, 105
amplitude, 17, 20
assessment, 20, 129, 148
assets, 3, 39, 128, 129, 132, 143, 152, 154, 162, 167, 168, 169, 173, 193
attacks, vii, viii, 127, 129, 134, 135, 142, 144, 146, 150, 172
auditing, 131, 173
authority, 156, 158, 168, 192, 194
authors, 39, 97, 99, 122, 126
availability, 3, 39, 172, 187
averaging, 112, 113, 119
awareness, 132, 141, 193

B

background, vii, 2, 3, 4, 5, 6, 8, 16, 18, 19, 21, 27, 29, 37, 38, 39, 41, 105, 106, 107, 121, 174
background information, 174
background noise, vii, 2, 3, 4, 5, 6, 8, 16, 18, 19, 21, 27, 29, 37, 38, 39, 41, 105, 106, 107
bandwidth, 17, 148
banks, 102
barriers, 137
beams, 196
behavior, 5, 6
benchmarks, 160, 169, 195
bias, 17
blind spot, 128, 132, 141, 142
blocks, 20
breathing, 2
broadband, 94, 129, 131, 133, 148, 149, 157, 158, 159, 161, 192, 193, 195
building code, 177
business model, 159

C

cables, 138
carrier, 16, 22, 93, 138
case study, 130, 131, 135, 138, 143, 144, 150, 153, 154, 163, 169, 170, 172, 173, 174
cell, 142, 185, 193
certification, 154
channels, 103, 163, 195
children, 103
clarity, 112
classification, 97, 126
coastal communities, 177
codes, 177
coding, 4, 5, 6, 7, 17, 18, 101, 106, 123
coffee, 121
collaboration, 128, 130, 131, 132, 133, 134, 137, 151, 155, 156, 157, 166, 167, 168, 170, 173
commerce, 181
communication systems, vii, 2, 4, 5, 6, 7, 8, 9, 23, 37, 38, 39, 100, 101, 105, 129, 171
community, 97, 136, 187, 194
complexity, 102, 122, 157
compliance, 94, 154, 194
components, 145, 157, 166, 184
composition, 196
compounds, 132
computation, 9, 19, 24, 98
concrete, 183
confidence interval, 25, 112
configuration, 18, 19, 22
confusion, 116, 163
consensus, 168
construction, 192
contamination, 179

continuity, 128, 131, 138, 141, 153, 154, 167, 192
control, 5, 110, 121, 122, 161, 175, 184, 185
copper, 185
costs, 159, 181
covering, 150
cycles, 144
cyclones, 177

D

data analysis, 4
data collection, 111
database, 5, 16, 17, 102, 103, 110, 122
deaths, 135, 144, 146, 177, 193
decision makers, 163
decoding, 105
defense, 137, 154
definition, 19, 24, 134, 160
degradation, 3, 39
delivery, vii, 21, 100, 101
Department of Homeland Security, viii, 39, 122, 127, 128, 129, 171, 188, 192, 193, 196
Department of Justice, 129, 131
Department of the Interior, 191, 196
deposits, 181
destruction, 177
detachment, 154
detection, viii, 100, 101, 103, 105, 106, 107, 112, 118, 119, 120, 121, 183
detonation, 172, 178, 179, 180
deviation, 23, 24, 104
diamonds, 117
digital communication, 38, 105
Digital Problem Working Group, vii, 2, 4
digital television, 136, 157
discipline, 22
disclosure, 5
disseminate, 136
distortions, 101, 121
distribution, 23, 97, 114, 126, 170, 180
diversity, 130, 172, 193
draft, 133, 165, 168, 170
drainage, 175
drinking water, 174
duplication, 156, 167
duration, 3, 104, 122
duties, 163

E

earth, 184, 193
emergency management, 136
emergency planning, 162
emergency preparedness, 136, 138, 153, 187
emergency response, 134, 136, 147, 148, 154, 159, 167, 170, 192
emotional state, vii, 100, 101
emotions, 101, 103
encoding, 17, 19, 22, 105
encryption, 160, 163, 195
energy, 181, 193
engagement, 195
environment, 3, 4, 5, 8, 17, 18, 19, 23, 26, 27, 28, 29, 30, 31, 32, 33, 34, 35, 36, 37, 38, 121, 171, 192
evacuation, 178
excitation, 105
exercise, 153, 159, 194
expertise, 127, 133, 148, 167, 168, 187, 195
exposure, 107

F

failure, 135, 156, 167, 175, 185
family, 141
Federal Communications Commission, viii, 127, 129, 167, 168, 184, 195, 196
federal funds, 150
feedback, 107, 108, 137, 153, 163
feet, 177, 178
females, 115, 116
fidelity, 21
fire hazard, 170
fires, 144, 179
flank, 184
flood, 130, 171, 174, 175, 180
flooding, 138, 141, 174, 175, 177, 185
freedom, 1, 23
frustration, 146
fuel, 140, 180
funding, 128, 129, 132, 144, 148, 150, 151, 160, 173, 194
funds, 150, 151, 152, 159, 194

G

gender, 20, 21, 98
generation, 18, 178, 180
geography, 142, 172
goals, 128, 133, 137, 147, 150, 151, 155, 156, 157, 158, 161, 163, 194, 195, 196
governance, 147, 148, 149, 151, 152, 154, 155, 156, 160, 167, 194
grants, 128, 132, 136, 146, 150, 156, 173
graph, 26

gravity, 184
groundwater, 184
grouping, 94
groups, 20, 23, 24, 102, 133, 146, 148, 150, 151, 159, 163, 164, 165, 166, 167, 168, 173
growth, 187
guessing, 23, 114, 118
guidance, 128, 129, 132, 133, 146, 156, 162, 168, 173, 187, 192
guidelines, 151
guiding principles, 193

H

hazardous materials, 180
hazards, 134, 135, 138, 171, 172, 174, 179, 181, 194, 196
health, 179, 187
health care, 187
hearing loss, 106
heat, 179
height, 178
histogram, 104
homeland security, 172, 187, 194
hospitals, 180
host, 121, 122
house, 166, 194
housing, 181
hub, 184
hurricanes, 134, 135, 172, 176, 177, 196
hypothesis, 23

I

ideal, 8, 27
identification, vii, 100, 101, 102, 105, 108, 109, 110, 111, 114, 115, 116, 117, 122, 123, 169, 194
identity, 100, 108, 109
images, 192
implementation, 5, 6, 23, 100, 128, 133, 154, 157, 167, 169
Improved Multi-Band Excitation, vii, 1, 2, 4
inclusion, 3, 6
incompatibility, 132, 144
indication, 6, 9, 104, 115
industry, 97, 126, 131, 169, 173, 187, 194
information sharing, 147, 150
infrastructure, vii, 128, 129, 131, 134, 138, 142, 154, 159, 162, 167, 171, 172, 175, 180, 181, 184, 187, 192, 195
initial state, 109
innovation, 129
insertion, 105
instructional materials, 97, 126
instruments, 156
integration, 136, 187
interaction, 122, 150
interagency committees, 136
interface, 106, 108, 109, 110, 111, 185
internet, 143
interval, 25, 94
investment, 150, 152
isolation, 93, 144, 196

J

jurisdiction, 148, 159, 161

L

land, vii, 2, 4, 5, 6, 178, 184
language, 115
law enforcement, vii, 129, 132, 134, 137, 144, 154, 169, 172, 173, 193, 194
leadership, 136, 155, 156
legislation, 136, 166
levees, 175
licenses, 136, 148, 160
life expectancy, 145
lifespan, 183
lifetime, 171
light trucks, 142
likelihood, 1, 114, 118, 130, 171, 172, 179
limitation, 118
line, 26, 112, 113, 120, 186, 196
links, 187
listening, vii, 4, 5, 6, 7, 17, 19, 20, 21, 22, 24, 27, 29, 31, 33, 34, 35, 36, 38, 94, 100, 106, 107, 111, 121
listening experiments, vii, 100
local authorities, 150, 194
local government, 136, 150, 152
Louisiana, 138

M

maintenance, 159
males, 102, 115, 116
management, 131, 136, 153, 165, 173
manufacturer, 8, 154
manufacturing, 180
mapping, 144
matrix, 116
measurement, 17, 40, 98

measures, 157, 167, 168
media, 148, 187
median, 102, 103
membership, 165, 193, 196
memory, 102, 108
messages, 103, 104, 112, 113, 119
meteor, 193
missions, 148, 156, 162
mixing, 105
mobile communication, 143
mobility, 185
model, 1, 105, 114, 171
modules, 185
money, 159
morale, 171, 192
movement, 177, 193

N

naming, 163
nation, 136, 142, 177
National Park Service, 139, 154, 163
National Response Framework, 171, 192, 196
national security, 97, 126, 153, 187, 192
natural disasters, 130, 135, 175
natural gas, 181
natural hazards, 130
NCS, 129, 153, 165
negotiating, 158, 161
noise, 2, 3, 4, 5, 6, 8, 9, 17, 18, 19, 20, 21, 22, 23, 26, 27, 29, 31, 33, 34, 35, 36, 37, 38, 39, 94, 102, 103, 105, 123
NTIA, 97, 99, 126, 129, 136, 137, 151, 152, 170, 192, 195
null hypothesis, 23

O

objectives, 131, 147, 155, 169, 173, 194, 195, 196
obligation, 195
oil, 181
openness, 193
order, 7, 17, 19, 20, 94, 106, 109, 136, 148, 152, 160, 166, 195
overload, 19, 93, 102
oversight, 136

P

parameters, 1, 9, 105, 121, 186
partnership, 148, 195
PCM, 2, 99, 105
peer review, 152
perceptions, 173
permit, 195
photographs, 132, 143
pitch, 101, 102, 103, 104, 105, 116
planning, 3, 39, 128, 129, 132, 133, 134, 144, 147, 151, 152, 153, 160, 162, 167, 172, 173, 194, 195
police, 131, 146, 150, 194
policy initiative, 194
poor, 163
population, 23, 134, 171, 172, 192, 196
portraits, 110, 126
power, 2, 8, 17, 19, 94, 139, 176, 177, 180, 184, 185
preparedness, 134, 136, 137, 148, 150, 159, 162, 163, 172, 187, 193, 194
presentation order, 17, 20, 21
pressure, 98
prevention, 194
private sector, viii, 127, 129, 142, 147, 156
probability, 114, 117, 119, 174, 181
production, 17, 18, 123
program, 108, 131, 136, 151, 152, 153, 154, 172, 173, 187, 192, 194
PSD, 10, 11, 12, 13, 14, 15
public education, 193
public health, 134, 187
public service, 184
pulse, 180

R

radiation, 179, 180
radio, vii, 2, 3, 4, 5, 6, 8, 18, 19, 37, 39, 129, 132, 136, 138, 142, 144, 150, 154, 163, 184, 185, 186, 193, 196
rainfall, 174, 177
range, viii, 9, 17, 21, 94, 100, 102, 103, 105, 106, 107, 109, 110, 111, 119, 128, 132, 142, 146, 157, 169, 172, 180, 184, 185, 194
reading, 102
realism, 122
recognition, 101, 103, 122, 123
recovery, 134, 136, 148, 193, 194
region, 94, 141, 142, 163, 172, 179, 181
relationship, 121, 157, 158, 160, 168, 195
relevance, 133, 165
reliability, 148, 180, 187, 192
repair, 141
resource allocation, 193
resources, 128, 130, 133, 134, 135, 137, 155, 156, 162, 165, 167, 168, 173, 192, 194
revenue, 172

rhythm, 121
risk, 136, 150, 153, 156, 176, 177, 178, 179, 194
robustness, 105, 113, 121
rolling, 153
runoff, 175

S

safety, 2, 3, 4, 38, 39, 100, 103, 133, 135, 136, 144, 148, 151, 157, 159, 180, 192, 193, 195
satellite, 143, 184
saturation, 17, 120, 184
school, 180
scores, 23, 26, 27, 28, 29, 30, 31, 32, 33, 34, 35, 36, 37, 50
sea level, 177
search, 102, 137, 171
secondary information, vii, 100, 101
security, 192, 195
sediment, 139
selecting, 102, 130, 166, 171
service provider, 185
shape, 9, 113
sharing, 135, 149, 158, 161
shelter, 141
signals, 19, 94, 103, 105, 142, 184, 186, 196
signal-to-noise ratio, 8
signs, 177
skills, 159
software, 5, 9, 18, 93, 101, 105, 109, 110, 122, 151, 186
soil, 181
space, 184
spectrum, 1, 3, 39, 94, 98, 123, 129, 133, 136, 148, 149, 157, 158, 161, 184, 186, 192, 193, 195
spectrum allocation, 148
speech processing, viii, 8, 100, 101, 103, 105, 107, 112, 113, 114, 116, 117, 119, 120, 121
speed, 176, 178
stakeholder groups, 132, 133, 148, 163, 164, 165, 166, 168, 169, 172, 173
stakeholders, 138, 150, 151, 153, 156, 158, 166, 170, 195
standard deviation, 23, 24, 104
standardization, 123
standards, 94, 105, 129, 131, 154, 172, 173, 194, 195
stars, 117
stimulus, 122
storage, 180
storms, 176
strategic planning, 151
strategies, 128, 133, 137, 155, 156, 157, 161, 167, 169, 173, 194
strength, 102, 105
stress, 100, 103, 112, 122, 123
subscribers, 153
supervision, 39, 122
suppliers, 154
supply, 94, 157, 184
support services, 194

T

tanks, 140
targets, 194
task performance, 112
technical assistance, 128, 133, 153, 156, 162, 168
telecommunications, 131, 138, 142, 148, 153, 161, 169, 173, 187, 192, 196
telecommunications services, 153
teleconferencing, 143
telephone, 22, 123, 139, 153, 186, 194, 196
television, 136, 143
tension, 146
territory, 151
terrorism, 163, 172, 192
test data, 97, 126
test procedure, 17
thinking, 18
threat, 162, 183
threats, 134, 137, 179, 184
threshold, 19
tides, 177
tornadoes, 135, 139
tracking, 103, 164, 165, 193
tracks, 164, 165
trade, 131, 173
training, 8, 18, 98, 101, 103, 107, 108, 115, 116, 122, 144, 145, 151, 152, 153, 159, 160, 194
transformation, 117, 118
transition, 136, 157
translation, 1
transmission, 5, 18, 27, 29, 31, 34, 35, 36, 37, 39, 40, 100, 105, 142, 144, 184, 192
transparency, 166, 196
transportation, 141, 180, 181
trees, 177, 193
tremor, 181
trial, 107, 118
triggers, 178, 184

U

uncertainty, 146
uniform, 94

urban areas, 179

V

variability, 21
variance, 23
vehicles, 143, 154, 184
victims, 135
vision, 133, 157, 167, 169
vocoder, vii, 2, 4, 5, 6, 8, 23, 24, 38, 101
voice, vii, 2, 3, 4, 18, 19, 20, 22, 36, 100, 101, 105, 116, 121, 123, 129, 141, 143, 150, 154, 159, 167, 186, 192
voicing, 102, 105
vulnerability, 128, 131, 138, 139, 162, 192

W

wear, 38
wind, 141, 176
windows, 177
winning, 148, 149, 158, 192, 193
wireless systems, 193
wood, 3, 6, 8, 9, 10, 37
working groups, 149, 151, 157, 159, 166, 194
writing, 193